Composites Science and Technology

Series Editor

Mohammad Jawaid, Laboratory of Biocomposite Technology, Universiti Putra Malaysia, INTROP, Serdang, Malaysia

This book series publishes cutting edge research monographs comprehensively covering topics in the field of composite science and technology. The books in this series are edited or authored by top researchers and professional across the globe. The series aims at publishing state-of-the-art research and development in areas including, but not limited to:

- Conventional Composites from natural and synthetic fibers
- Advanced Composites from natural and synthetic fibers
- Chemistry and biology of Composites and Biocomposites
- Fatigue damage modelling of Composites and Biocomposites
- Failure Analysis of Composites and Biocomposites
- Structural Health Monitoring of Composites and Biocomposites
- Durability of Composites and Biocomposites
- Biodegradability of Composites and Biocomposites
- Thermal properties of Composites and Biocomposites
- Flammability of Composites and Biocomposites
- Tribology of Composites and Biocomposites
- Applications of Composites and Biocomposites

Review Process

The proposal for each volume is reviewed by the main editor and/or the advisory board. The chapters in each volume are individually reviewed single blind by expert reviewers (at least two reviews per chapter) and the main editor.

Ethics Statement for this series can be found in the Springer standard guidelines here - https://www.springer.com/us/authors-editors/journal-author/journal-author-helpdesk/before-you-start/before-you-start/1330#c14214

Mohammad Jawaid · Anish Khan
Editors

Multifunctional Boron-Nitride Composites

Preparation, Properties and Applications

Springer

Editors
Mohammad Jawaid
Laboratory of Biocomposite Technology
INTROP, Universiti Putra Malaysia
Serdang, Selangor, Malaysia

Anish Khan
Center of Excellence for Advanced
Materials Research
King Abdulaziz University
Jeddah, Saudi Arabia

ISSN 2662-1819 ISSN 2662-1827 (electronic)
Composites Science and Technology
ISBN 978-981-99-2865-1 ISBN 978-981-99-2866-8 (eBook)
https://doi.org/10.1007/978-981-99-2866-8

This Springer imprint is published by the registered company Springer Nature Singapore Pte Ltd.
The registered company address is: 152 Beach Road, #21-01/04 Gateway East, Singapore 189721, Singapore

Contents

BN-Based PCM Composites for Thermal Management: Synthesis and Performance Assessment 1
Usman Bin Shahid, Mohammad Owais, Muhammad Humza Javed, and Ahmed Abdala

Synthesis of Heteroatoms Including at Least One Boron Metal Complexes and Their Catalytic Applications 47
Md. Zafer Alam and Salman A. Khan

Thermally Conductive Hexagonal Boron Nitride-Polyethylene Nanocomposites: Effect of Processing Method and Filler Size, Exfoliation, and Alignment on Thermal Conductivity 73
Mehamed Ali and Ahmed Abdala

Thermal Conductive Composites Reinforced with Advanced Micro and Nano-sized Boron Nitride Particles 93
Alok Agrawal and Alok Satapathy

Graphene-Analog Boron Nitride Nanomaterial and Their Photocatalytic Applications 115
Alimuddin and Salman A. Khan

An Extensive Study on Thermo-Mechanical Characterization of Hexagonal-Boron Nitride Polyester Composites 131
Debasmita Mishra and Anish Khan

Luminescence of Boron Nitrides 149
Young-Kuk Kim and Jae-Yong Jung

Boron Nitride Nanocomposites Used as Hydrogen Storage Material 163
Md. Mohasin, Md. Zafer Alam, Qasim Ullah, and Salman A. Khan

About the Editors

Dr. Mohammad Jawaid is currently working as Senior Fellow (Professor) at Biocomposite Technology Laboratory, Institute of Tropical Forestry and Forest Products (INTROP), Universiti Putra Malaysia, Serdang, Selangor, Malaysia, and also Distinguish Visiting Professor at MJIIT-UTM, Malaysia. So far, he has published 70 books, 85 book chapters, more than 450 peer-reviewed international journal papers, and several published review papers under top highly cited articles in Science Direct. He also obtained 6 Patents and 6 Copyrights. H-index and citation in Scopus are 79 and 27691 and in Google scholar, H-index and citation are 92 and 37002. He worked as Chief Executive Editor of Pertanika UPM Journals from 2021–22. He is founding Series Editor of Springer-Nature-3 Book Series: Composite Science and Technology; Sustainable Materials and Technology; and Smart Nanomaterials and Technology, and also Series Editor of Springer Proceedings in Materials, and also International Advisory board member of Springer Series on Polymer and Composite Materials. Presently he working as guest editor of special issues of Industrial Crops and Products and Journal of Renewable Materials. He is also Editorial Board Member of Journal of Polymers and The Environment, Journal of Natural fibres, Journal of Plastics Technology, Applied Science and Engineering Progress Journal, Journal of Asian Science, Technology and Innovation and the Recent Innovations in Chemical Engineering. He has more than 20 years of experience in teaching, research, and industries. His area of research interests includes hybrid reinforced/filled polymer composites and advanced materials. His five published review papers under hot-cited articles in science direct during 2016–2019. He is Reviewer of several high-impact ISI journals (200 journals). He also delivered plenary and invited talks in international conferences related to composites in India, Turkey, Malaysia, Thailand, United Kingdom, France, Saudi Arabia, Egypt, Hongkong, and China. Besides that, he is also a member of technical committees of several national and international conferences on composites and material science. Dr. Mohammad Jawaid received Excellent Academic Award in Category of International Grant-Universiti Putra Malaysia-2018 and also Excellent Academic Staff Award in industry High Impact Network (ICAN 2019) Award. Beside that Gold Medal-Community and Industry Network (JINM Showcase) at Universiti Putra Malaysia. He also received Publons Peer Review Awards 2017, and 2018

(Materials Science), Certified Sentinel of science Award Receipient-2016 (Materials Science) and 2019 (Materials Science and Cross field). Recently he received Silver Medal in International Technology Expo (ITEX-22), Malaysian Technology Expo (MTE-21), and MTE-SDG International Innovation Awards 2021. He is also Winner of Newton-Ungku Omar Coordination Fund: UK-Malaysia Research and Innovation Bridges Competition 2015. Recently he recognized with Fellow and Charted Scientist Award from Institute of Materials, Minerals and Mining (IOM), UK. He is also Vice President and life member of Asian Polymer Association (APA), and life member of Malaysian Society for Engineering and Technology (MySET). He has professional membership of American Chemical Society (ACS), and Society for polymers Engineers (SPE), USA.

Dr. Anish Khan is an Associate Professor, at the Centre of Excellence for Advanced Materials Research (CEAMR), Faculty of Science, King Abdulaziz University, Jeddah, Saudi Arabia. He received his Ph.D. from Aligarh Muslim University, India, in 2010. He has research experience working in the field of synthetic polymers and organic–inorganic electrically conducting nano-composites. He completed a Post-doctoral in Electroanalytical Chemistry from the School of Chemical Sciences, Universiti Sains Malaysia (USM) in 2010–2011. He has research and teaching experience and published over 300 research papers in referred international journals. He attended more than 20 international conferences/workshops and has published 60 books, six in progress, and 107 book chapters. He has completed around 31 research projects. He served as an Editor and board member of the *Journal of Natural Fibers* (JNF) and a Member of the American Nano Society. Published two US patents. His fields of specialization are polymer nano-composite/cation-exchanger/chemical sensor/micro biosensor/nanotechnology, application of nano-materials in electroanalytical chemistry, material chemistry, ion-exchange chromatography, and electroanalytical chemistry, dealing with the synthesis, characterization (using different analytical techniques) and derivatization of inorganic ion-exchanger by the incorporation of electrically conducting polymers, preparation, and characterization of hybrid nano-composite materials and their applications, polymeric inorganic cation-exchange materials, electrically conducting polymeric, materials, composite material use as sensors, Green chemistry by remediation of pollution, heavy metal ion selective membrane electrode, and biosensor on the neurotransmitter.

BN-Based PCM Composites for Thermal Management: Synthesis and Performance Assessment

Usman Bin Shahid, Mohammad Owais, Muhammad Humza Javed, and Ahmed Abdala

Abstract Boron nitride (BN), a 2D material analogous to graphene with highly desirable physical and chemical properties has captivated researchers for its immense potential in thermally enhanced composites. This chapter briefly discusses the history of BN and its evolution as a novel nanomaterial, and its application in thermal management systems, followed by a detailed overview of various synthesis methods used to synthesize these composites and strategies to improve and enhance the overall composite performance. We also discuss the figure-of-merit approach to assess the performance of these BN-based composites concluding with a discussion on the way forward.

Keywords Phase change materials · Boron nitride · Composite · Performance metric · Figure of merit · Thermal management

U. B. Shahid & M. Owais: Authors contributed equally.

U. B. Shahid (✉)
Division of Sustainable Development, College of Science and Engineering, Hamad Bin Khalifa University, Education City, Qatar Foundation, PO Box 5825, Doha, Qatar
e-mail: ubshahid@connect.ust.hk

Department of Chemical and Biological Engineering, Hong Kong University of Science and Technology, Clear Water Bay, Kowloon, Hong Kong, China

M. Owais
Center for Materials Technologies, Skolkovo Institute of Science and Technology, 121205 Moscow, Russia
e-mail: mohammad.owais@skoltech.ru

M. H. Javed
Department of Mechanical and Aerospace Engineering, Hong Kong University of Science and Technology, Clear Water Bay, Kowloon, Hong Kong, China
e-mail: mhjaved@connect.ust.hk

A. Abdala
Department of Chemical Engineering, Texas A&M University at Qatar, Education City, PO Box 23874, Doha, Qatar
e-mail: ahmed.abdala@qatar.tamu.edu

M. Jawaid and A. Khan (eds.), *Multifunctional Boron-Nitride Composites*, Composites Science and Technology, https://doi.org/10.1007/978-981-99-2866-8_1

1 Introduction

By definition, a 'composite material' is a combination of two or more components forming two or more distinct phases with recognizable interfaces [1]. Combining two materials at an optimum ratio into a composite can result in properties that transcend the properties of the individual components. While the concept of a 'composite' material may appear to be somewhat new or recent; however, strictly speaking, this is not the case. Nature is replete with marvelous examples of composite materials. Wood, for instance, in principle is a fibrous composite: a lignin matrix supporting cellulose fibers. While the cellulose fibers boast high tensile strength, this uniqueness is blemished by their low stiffness (i.e., high flexibility), which is furnished by the lignin matrix as it entangles the fibers in a rigid formation. Similarly, the human bone is another spectacular example of a composite material: a mineral matrix, called 'Bone apatite', supporting the soft and short collagen fibers resulting in outstanding skeletal support for the human muscle mass [2].

The ultimate purpose of composite materials is to optimize material properties, including chemical, mechanical, and physical properties, to meet the application requirement. Physical properties of common interest include mechanical, thermal, electrical, optical, and acoustical properties. The Third Industrial Revolution (the 1970s), driven by the rapid advancements in electronics and computers, digitization, and automation, led to increased demand for materials with superior performance. While a monolithic engineering material that satisfies such a need would have been highly welcome, such materials are hardly ever available. This inevitable need fueled the demand for novel and superior composite materials. A typical composite material comprises two phases: (a) the matrix (or the binder) and (b) the filler. The matrix material is dominant in the final composite, while the filler is preferably kept marginally above the percolation threshold. The filler (often the reinforcing material) is embedded in the matrix to produce the desired amplification of properties. The filler must be well dispersed and distributed to ensure that the property enhancement is pronounced and uniform.

The recent advent of novel nanomaterials has drastically revolutionized the field of composite materials. More specifically, the discovery of graphene smashed the realm of possibilities. It fueled a new era of nanotechnology [3, 4], spurring experimental and theoretical research for other possible materials with a similar 2D structure as graphene. We now have a plethora of similar materials, among which boron nitride (BN) has recently become the center of attention. Boron neighbors the carbon atom in the periodic table of chemical elements and similarly can form an extended network of covalent bonds. Interestingly, boron can also imitate the same 2D-hexagonal structure of graphene, albeit with the substitution of the carbon atoms with alternating B and N atoms [3, 5]. This typical hexagonal structure (hBN) has a lot in common with graphene; however, subtle differences in the chemistry of hBN bonds make all the difference.

1.1 History of BN

Elemental Boron having a black metal-like appearance was reported as early as 1808 by Gay-Lussac and Thenard [6]. However, the Arabs used borax, the most popular boron derivative, in melting processes as early as the sixteenth century. Boron is a relatively rare element, with an estimated mass abundance of only 0.001%. However, concentrated deposits of boron minerals are easily found around the globe, with major reserves in Italy, Chile, the United States of America, Argentina, Peru, and Bolivia [6].

Notwithstanding the widespread availability of boron derivatives, their chemistry remained relatively unexplored until the early nineteenth century. This hindrance is primarily attributed to the inherent difficulty of handling many of these compounds alongside the facile degradation of many boron compounds subjected to hydrolysis or oxidation. Nevertheless, it was the pioneering work by Stock around 1905 that genuinely explored the chemistry behind boron as he exhaustively investigated boron hydrides. However, boron-nitrogen compounds date earlier than Stock's investigation of boron-hydrides. The earliest report on BN synthesis was by Balmain in 1842 [7], but it was a century later that its true potential would emerge in the commercial market. Detailed investigation on boron-nitrogen compounds was carried out much later in 1926 when Stock and Pohland reported their discovery of borazine $(-BH-NH-)_3$. Borazine, dubbed the "inorganic benzene", was the product of the reaction between diborane and ammonia and is considered a significant breakthrough in boron-nitrogen chemistry. However, research interest did not really pick up until around the 1950s, when the scientific community really started developing this research front, driven purely by research interest and not commercial incentives. However, Wentorf's [8] discovery of the 2nd hardest material (after diamond), the cubic-BN, in 1957 put BN on the scientific community's radar, and research interest ensued. BN's potential as a ceramic material further fueled research interest, and it became commercially available by 1969 [9] (Fig. 1).

As a neighbor of carbon in the periodic table, Boron also offers chemistry similar to carbon's, e.g., the isoelectric nature of bonds in both B–N and C–C. Although the sum of atomic radii of B and N is similar to that of the two C atoms, making

Year	Researcher	Contribution
1809	Gay-Lussac	Adduct of ammonia and trifluoro-borane
1842	Balmain	First reports on boron nitride
1850	Wohler, Rose	Characterization of boron nitride
~1905	Stock, Joannis	Studies on the interaction of ammonia with trihalogenoboranes
1926	Stock and Pohland	Synthesis of borazine
1935	Wiberg, Schlesinger	Fundamental studies of boron-nitrogen compounds
1955	Brown and Laubengayer	Preparation of B-trichloroborazine with standard laboratory equipment
1956	Dewar and coworkers	Heteroaromatic boron-nitrogen compounds
1958	Parry and coworkers	Structure of diammoniate of diborane

Fig. 1 Timeline of breakthroughs for Boron-Nitrogen chemistry

BN compounds similar to carbon-based materials (e.g., graphite, diamond, etc.) in some aspects, such comparison cannot be excessively pursued. Despite their isoelectric nature and similar bond lengths, they are not identical. Boron and nitrogen are two completely different elements, contrary to the case of the C–C bond, and their differing electronegativity results in an altered bond symmetry. These subtle differences certainly outweigh their similarities to their inorganic counterparts. Notwithstanding the remarkable properties and unique chemistry B-N compounds have to offer, the one that has most intrigued researchers is its 2D allotrope—Boron nitride—the inorganic analogy to graphene.

1.2 Boron Nitride (BN) Properties

Boron nitride is a soft white crystalline material (often called "white graphite"), essentially comprising a network of covalently bonded B and N atoms (typical morphologies of BN powders are shown in Fig. 2). Akin to carbon, BN can also exist in several different phases, with the most common ones including; (a) the soft hexagonal layered structure (h-BN) (similar to graphene), (b) the hard cubic structure (c-BN), and (c) super-hard hexagonal phase in wurtzite-type (w-BN). It must be noted that the BN planes in h-BN are stacked atop each other without any horizontal displacement, however, with the boron and nitrogen atoms aligned along the c-axis (Fig. 3). With the higher electronegativity of nitrogen atoms, the π-electrons are restricted at the nitrogen atom and render the entire compound electrically insulative, contrary to the case of graphite. The most common and stable form among the three allotropes is the h-BN, whereas w-BN is one of the most unstable and rare BN polymorphs.

The B-N bond length of h-BN, i.e., 1.466 Å, is marginally larger than the graphite C–C bond length, i.e., 1.421 Å. This increased bond length is indicative of the electron resonance between boron and nitrogen atoms. Moreover, the lack of delocalized π-electrons results in a slightly lower inter-layer spacing of 3.331 Å in h-BN than the

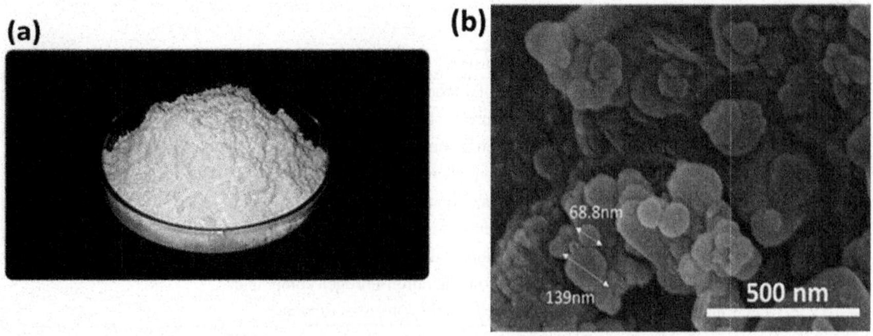

Fig. 2 **a** Pristine powder of h-BN; **b** SEM image of spherical h-BN

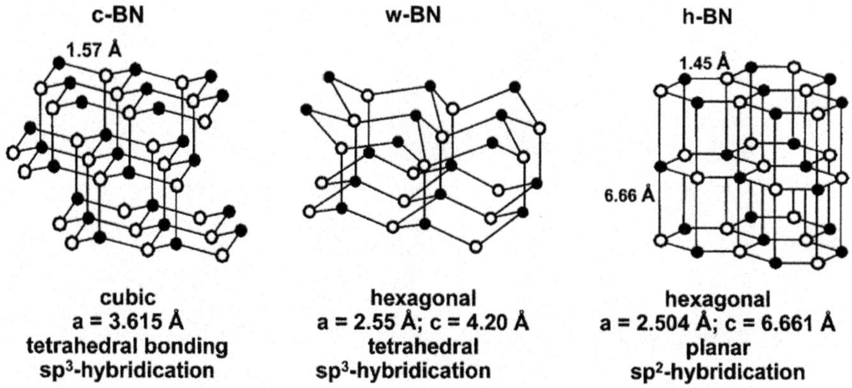

Fig. 3 Crystal structures of BN allotropes **a** cubic (c-BN), **b** wurtzitic (w-BN), and **c** hexagonal (h-BN). Reproduced with permission from Haubner et al. [10]. Copyright 2002 Springer Nature

~3.440 Å in graphite. This unique non-covalent nature of the B-N bond makes h-BN mechanically hard, while the layered structure with weak Van der Waals contributes to its high lubricity [11, 12]. This combination of hardness and lubricity with a low density (i.e., 1.9–2.3 g cm^{-3}) makes it a unique candidate for a wide variety of ceramic applications [13].

However, BN's exceptional physical properties elicited research interest, and its chemical stability and inertness give key advantages over its organic counterpart, i.e., graphene. While graphene and other carbon allotropes readily combust when exposed to air at high temperatures, BN is relatively inert to such exposure and is considered a flame-resistant material [14]. Even at extremely high temperatures, i.e., ~1000 °C, BN remains impervious to oxygen attack (see Fig. 4), while in inert environments, it can withstand exposure to temperatures as high as ~2800 °C. For this reason, h-BN as a ceramic material easily outperforms other top-performing ceramics like Si_3N_4, Al_2O_3, or SiC. Similarly, the temperature resistance of h-BN compares with that of MgO, ZrO_2, or CaO; however, BN showcases a higher thermal shock resistance than these oxides. BN's chemical nature also makes it impervious to wetting by most metallic (e.g., Al, Cu, Zn, Fe, and Ge) and non-metallic (e.g., Si, B, SiO_2, cryolite, and halides) melts. Detailed information about the chemical and physical properties of h-BN is available in the literature [15, 16].

In addition to these exceptional chemical and mechanical properties, BN has become even more famous for its outstanding thermal properties. As is the case for layered materials, they are often characterized by their anisotropic properties, primarily electrical and thermal conductivity. Graphene, for example, has a high in-plane thermal and electrical conductivity in contrast to its cross-plane numbers for the same. Similarly, BN is theoretically estimated to have a high in-plane thermal conductivity of ~550 W m^{-1} K^{-1} and a cross-plane conductivity of ~5 W m^{-1} K^{-1} at room temperature [17]. These high values for thermal conductivity put it in the same league as graphene and its derivatives—while not entirely at par but in

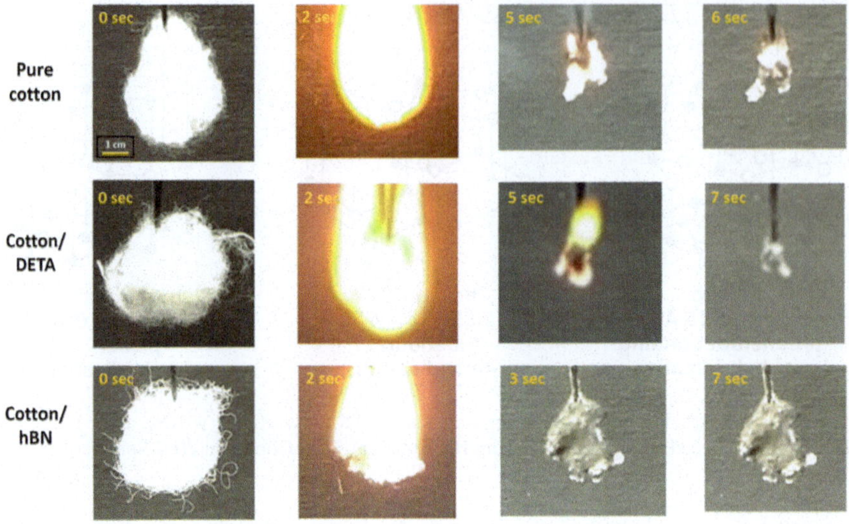

Fig. 4 Comparison of fire retardancy and self-extinguishing behavior of pure cotton, cotton/DETA, and hBN-coated cotton fabrics under single burning via vertical flame testing. Reprinted with permission from Ambekar et al. [14]. Copyright (2020) American Chemical Society

the competitive zone at least. The lack of electrical conductivity is not entirely a drawback; it can be considered a boon for applications where insulative properties are essential. Nevertheless, the high in-plane thermal conductivity, in addition to its atomic flatness, makes it an ideal filler candidate for composites with application in the thermal management of devices [18].

While graphene has been extensively studied for a vast range of potential applications, especially in applications where its 2D nature and exceptional thermal and electrical conductivity are envied upon, BN remains relatively unexplored. Experimental studies conducted thus far have only barely scratched the potential of BN since a measured thermal conductivity in the range of only 220–420 W m^{-1} K^{-1} has been reported thus far—a value well below the theoretical estimates of its properties. More assiduous investigations are still needed to explore its potential fully, especially in the field of composite materials, where its thermal conductivity and flame-resistivity make it a highly appealing candidate. This amalgamation of superior properties of h-BN thus widens its enormous range of technical applications in our daily life and industry. However, this chapter will focus mainly on thermally enhanced BN composites for thermal management applications.

2 Thermal Management

More recently, the research focus has switched to renewable energy systems due to the dire global warming concerns and increasing fossil fuel prices. Despite improved efficiencies of renewable systems, the research remains in its early phase. The challenges with the practical storage of this inconsistent renewable energy are still being worked on by scientists. Even though solar cells are at the forefront of research into renewable energy sources, having enough energy storage and recovery for a constant supply of energy remains a huge hurdle.

This problem for renewable energy systems can be effectively solved by thermal energy systems (TES). Utilizing a material's latent heat capacity, for instance, allows for the storage of significant amounts of heat energy in a relatively smaller volume with only a slight change in temperature within the media [19]. This concept was introduced by Telkes in the 1940s [20], but it only became the focus of extensive research in solar heating systems when the world went through the energy crisis of the 1970s [21]. Albeit when the energy crisis ended, the research interest in TES systems also dropped, however, the focus then moved to utilizing latent heat storage systems for waste heat recovery or load-leveling in power generation [22–24]. Nevertheless, recently, the focus of research has switched to TES systems, primarily as a result of the global push for a future powered by renewable energy where most industries are now being driven towards electrification [25, 26]. Research in this area is seeing a rise in the investigation of innovative multi-generation systems with PCMs as TES systems [25–28]. To optimize the thermal window for solar panel performance, PCMs hold great promise. The cells' temperature, which is a key factor in the photoelectric phenomenon that converts light into electricity, is essential for the overall performance efficiency of the panel [29].

Solar farms may be located in sunny areas, however, if the daily average temperature there exceeds 50 °C, performance is dramatically reduced. Many researchers [30, 31] have shown direct and indirect parallels for assessing PV efficacy at various operating temperatures, accounting for elements like system configuration, wind speed, and the type of cell. With rising cell temperature, the conversion efficiency of PVs fall as the drop in the open-circuit voltage (Voc), exceeds the minor increase in the short-circuit current (I_{sc}). Average cell efficiency drop ranges between 0.45 and 0.65% per °C rise in temperature [31, 32].

Active and passive cooling can be used to manage a device's temperature. The active cooling technique makes use of external means to increase the transfer of heat and includes the forced flow of gases or liquids, using fans or pumps where the fluid flow rate is determined by the heat removal rate. Active cooling is dependent on the supply of external energy sources., making this technique more costly and less favored. In contrast, the passive cooling approach increases natural convection and heat dissipation, without external input using heat pipes, a sink, and a spreader assembly that boosts conductive, convective, and radiation heat transmission into the TES medium. Thus, PCMs portray a significant function here as the energy carrier inside the various heat spreader or sink assemblies.

The growth and development of the present microelectronics industry are emphatically dependent on the fabrication of miniaturized and extremely thermally conductive components that can rapidly dissipate thermal energy. Consequently, these can be used in the applications associated with the computer, automobile, military, aerospace, and mobile phone industries. Intense overheating in electronic devices globally has been labeled the key factor responsible for electronic equipment failures. For instance, overheating can steer to high-degree malfunction, printed circuit board failure, or short circuit, in specific [33, 34]. The problems of electronic device breakdown due to the intense heat dissipation calls for a need to fabricate capable thermal interface materials (TIMs), which can dispel the thermal energies at a fast speed. Hence, with the ever-burgeoning expansion in the electronic industry, particularly in electronic gadgets, we must fabricate such substrates that can provide the benefits of high thermal conductivity, lifetime, and reliability.

2.1 *Thermal Interface Materials*

With a million-dollar net business, thermal interface materials (TIMs) are widely employed globally in the electronic industry and packaging systems. These TIMs are often positioned between the two solid surfaces of a heat source and a heat sink to enhance the thermal conduction over the interface. The key aim of TIMs is to fill the vacancies/defects/voids or remaining minor gaps between the two connecting surfaces to attain a better-improved contact area, which in turn improves the thermal conductivity of the system [35]. Figure 5 presents a typical illustration of TIMs. A TIM-deficient interface will have a highly concentrated heat energy flux, leading to a significant temperature reduction there. A TIM reduces the temperature drop over the interface by filling in the gaps and spaces.

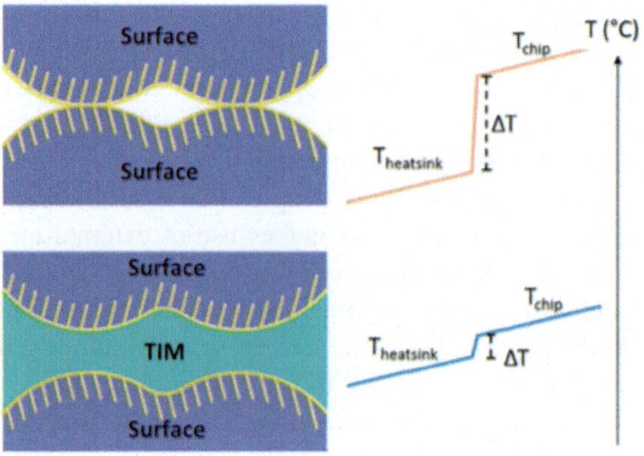

Fig. 5 Graphical illustration of the working principle of a TIM [36]

Presently commercially available TIMs are primarily divided into different groups; they are as follows:

- thermal pads,
- thermal grease,
- phase change materials,
- thermally conductive adhesives,
- thermally conductive papers/films
- gels and solders.

Although these materials have decent thermal conductivities, their use in electronics is severely constrained. For instance, when injected as a paste, thermal grease can be highly messy during processing and can pollute the working environment. Due to its liquidy paste-like shape, the thermally conductive layer can thin itself when forced, squeezing out from the edges, creating a dry interface and lowering the electronic system's overall performance. Henceforth, thermal grease is rarely selected as a strong contender for TIMs. Thermal pads, instead, are much easier to be used for widespread applications but are not long-lasting and reusable. Also, they are not as efficient in dissipating the heat generated by the electronic heat source as thermal pastes. Industrial research scientists are trying to improve their commercial thermal pads with typical thermal conductivity of over ~4–5 W/mK with simultaneous improvement in the characteristics of pads. Primary objectives include filling in any vacancy or void, containing high filler loading for attaining high thermal conductivity, and possessing higher elasticity and stiffness.

Thermally conductive thin films/papers are emerging as a new class of TIMs for future miniaturized electronics. They are predicted to improve the heat dissipation problem tremendously and reshape the thermal-management-based applications and thermal interface materials in the years to come [37–39]. These paper-like thermally conductive composite films reported in the literature can be classified into three categories as follows:

i. Polymer composites or nanocomposites,
ii. Carbonaceous materials with/without polymers,
iii. Carbon and ceramic-based composite materials.

These paper-like films are therefore an improved version of traditional heat spreaders or thermal interface materials with the thickness ranging from mm to μm. To attain the best thermal conductivity of the papers/films, factors like particle size, alignment of fillers material inside polymer, heat treatment process, mechanical pressing, and phenomena like interfacial phonon scattering of fillers in polymers should be actively pursued.

2.2 Thermal Energy Storage

Global energy consumption is on a rapid rise and current projections suggest this number will grow by 50% in 2050 [40]. Fossil fuels depleting rapidly, and their damaging impact on the environment has led researchers to think about a sustainable future. Thermal energy storage (TES) is the key to the success of any intermittent energy source in encountering demand. Especially in the case where the sun is scarcely available and solar energy applications are required.

In renewable energy systems, energy storage has become a vital part. In this thermal energy storage technology, thermal energy is supplied via heating or cooling a storage medium in which later the stored energy can be utilized, for example, in generating power. The main types of TES are shown in Fig. 6. An energy storage system can be described in the following characteristics [41]

- Storage period expresses how long the energy is stored
- Charging and discharging time refers to the time required to charge/discharge the system
- Capacity is the energy stored in the system
- Efficiency refers to the ratio of the energy provided to the energy needed to charge the storage system
- Power is how fast the energy is stored in the system
- Cost is the capacity or power of the storage system and its operational costs.

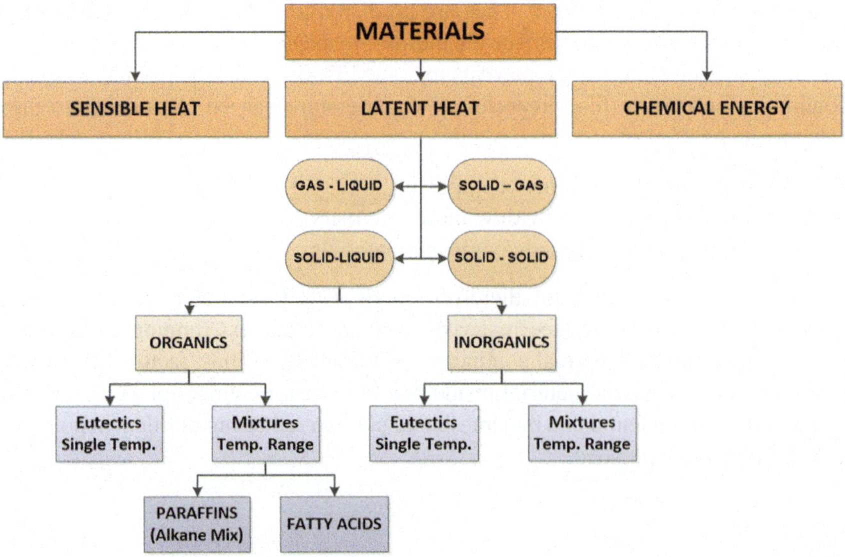

Fig. 6 Classification of energy storage materials. 'Reprinted from Ref. [42]. Copyright (1980), with permission from EDP Sciences'

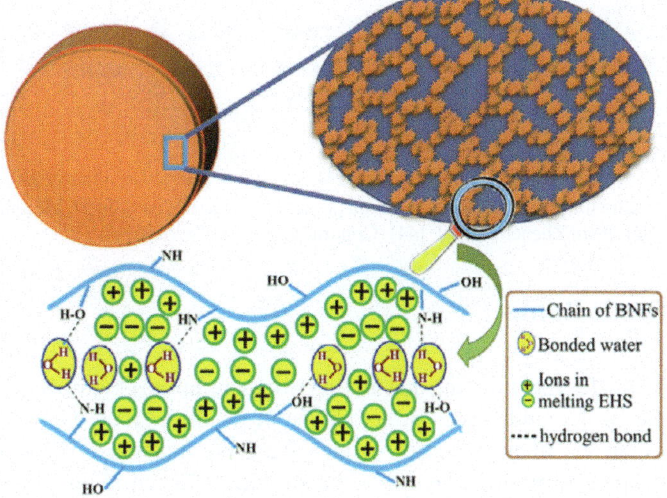

Fig. 7 EHS/BNF composite PCM developed by Weifang et al. [43]. Reprinted with permission from Weifang et al. [43]. Copyright (2019) Elsevier

Phase change materials (PCMs) are developed to enhance the performance of thermal energy storage systems. They store solar energy and industrial waste heat efficiently, especially when the PCM contains conductive fillers that enhance the thermal conductivity of PCMs. One such example is the boron nitride foam (BNFs) developed by Weifang et al. [43] as a packaging material for PCMs. The foam has high porosity (~97.6%) and low density (~1.7 mg/cm^3), offering adequate porosity for the adsorption of PCMs. Eutectic hydrate salt (EHS) was connected to BN foam via weak interactions. Moreover, the EHS/BNF composite PCMs displayed a low supercooling degree and high latent heat compared to conventional EHS/polymer. The thermal conductivity of EHS/4 wt.% BNF PCMs was 10.37 times the pristine PCMs and 2.24 times the EHS/polymer PCM (Fig. 7).

Zhenchao et al. [44] improved the thermal conductivity of pure paraffin by 600% by infiltrating the paraffin into an h-BN scaffold. This enhanced thermal conductivity shortened the phase change process, leading to more efficient energy storage and release. In addition, this 3D BN scaffold prevented the leakage of molten paraffin, indicating good shape stability of the overall PCM (Fig. 8).

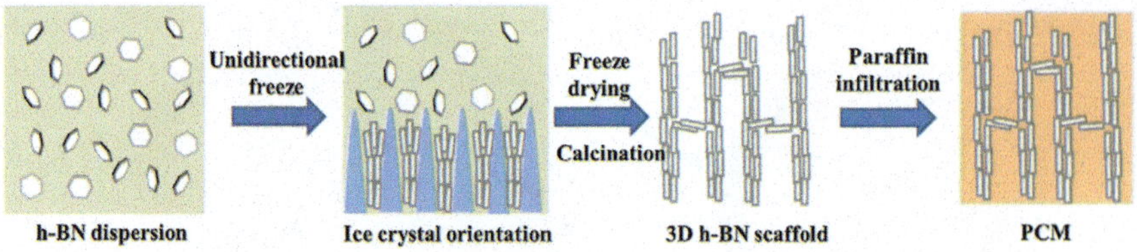

Fig. 8 Schematic illustration of the fabrication process of 3D h-BN/paraffin PCMs [44]. Reprinted with permission from Zhenchao et al. [44]. Copyright (2017) Elsevier

3 Synthesis Route

One may be misled into thinking that mixing two materials to obtain a new composite is as simple as it sounds. The correct choice of synthesis route plays a vital role in determining the final properties of the composite materials. The intrinsic and extrinsic properties of any composites are strongly affected by the entire fabrication process [11]. This section aims to provide a brief overview of these synthesis routes while explaining the fundamental principles behind these methods. Among the various techniques employed for preparing composite materials, the most popular and commonly resorted-to methods include the following:

1. Hot Pressing
2. Vacuum-assisted Filtration
3. Template Assembly
4. 3D-printing
5. Melt Impregnation
6. Vacuum Impregnation
7. Solution Impregnation.

While some techniques are suitable for thorough dispersion of the filler, others focus on maximizing the interaction between the filler and matrix by capitalizing on increased interfacial contact between the two. Other techniques that focus on improving the filler alignment are also available to either enhance or impair the anisotropic nature of properties. Therefore, it is vital to understand the purpose and philosophy of each technique, which is the primary objective of this section.

3.1 Hot Pressing

A common and popular method, the hot pressing technique, is often employed to prepare dense composites with an in-plane pie-pie stacking atoms orientation, significantly enhancing physical properties like thermal conductivity and mechanical strength [42, 45, 46]. This technique requires an external mechanical force which is usually engaged to assist in this process [47]. This composite densification process involves the concurrent application of pressure and temperature to the powder compact. The heating process is carried out via external heating elements, typically induction coils or graphite heating elements, whereas pressure is often hydraulic [46]. The densification mechanism can be split into three stages:

(i) *Particle rearrangement:* here open porosities are dramatically reduced while the closed porosity stays constant
(ii) *Plastic flow:* results in sealing of open porosity while the closed pores remain unaffected
(iii) *Volume diffusion and pore elimination:* results in the elimination of closed pores and increased densification.

Under pressure, the interfacial contact points amongst particles generate very high localized stress resulting in increased diffusion rates. The process parameters (i.e., temperature, pressure, heating rate, hold time) are all critical in determining the final properties and, as such, require careful optimization. In some cases, a controlled atmosphere may also be necessary whilst the densification process is carried out since elevated pressures and temperatures may trigger a reaction with ambient air, especially oxygen. Generally, the powder compact is heated to the maximum sintering temperature before applying pressure. Depending upon the requirement, the pressure may be increased in intervals.

The effectiveness of this method is remarkable, especially for composites with 2D nanofillers. As an example, Liu et al. [48] prepared a 3D segregated structure of polystyrene/polypropylene/boron nitride composites by employing a combination of the blend mixing method and the hot-pressing method. As shown in Fig. 1, the ternary-based poly-styrene/poly-propylene/boron nitride composites achieved a thermal conductivity of 5.57 W/(m·K) at a BN loading of 50 wt% [48]. The hot-pressing process facilitates the alignment of boron nitride platelets in conjunction with the polypropylene microspheres in the polystyrene polymer matrix to align in a 3D–segregated structure (Fig. 9). This unique morphological structure can then provide efficient heat conduction pathways for the transfer of phonons with an enhancement of mechanical and thermal properties.

Similarly, Liang et al. [49] used silane coupling agents for surface modification of BN sheets and aluminum nitride particles using the same blending/hot-pressing approach to attain composites with thermal conductivity of 2.60 W m^{-1} K^{-1} at a 40 vol% hybrid filler loadings. The surface treatment contributed to an extraordinary improvement in the interfacial contact between the fillers and the polymer, thus lowering the thermal interface resistance with more efficacious thermally conductive channels.

3.2 Vacuum-Assisted Filtration

Vacuum-assisted filtration is another effective method to create composites with aligned 2D fillers. This technique utilizes the force exerted by flowing water as it filters over the composite powder to align the 2D fillers in an in-plane orientation at the

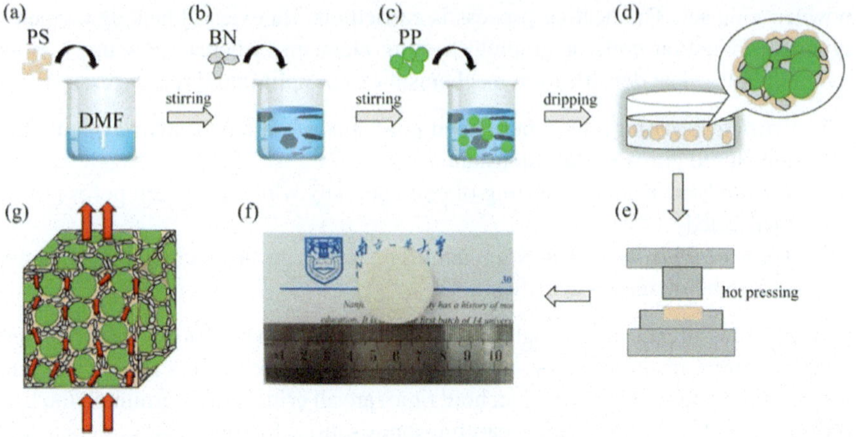

Fig. 9 The construction of the heat conduction pathways by the hot-pressing approach [48]. Reprinted with permission from Liu et al. [48]. Copyright (2019) Elsevier

interface of the filter paper. In this process, the liquid of the filler suspension is subsequently passed via pores of the membrane film, which normally involves pouring the solution onto filter paper with almost microscopic pores. At the filter's surface, a dense structure is created from the sedimenting filler material. A high-pressure vacuum is also created simultaneously to align the fillers and transfer phonons for high thermal conductivity. The filter's membrane or the paper is peeled off and dried in a heating oven after being cut to a specific thickness. The vacuum-assisted filtration process has, thus, been established to be tremendously rapid and facile for the fabrication of films or paper compared to the traditional polymer composite synthesis techniques used [35]. It is therefore simple to use and yields a high sheet alignment of fillers in the polymer. This natural procedure for the preparation of polymer nanocomposites has been extensively used by researchers in recent years [50, 51]. For instance, taking inspiration from nature's nacre papers, a non-covalent surface-modified boron nitride nano-sheets, and poly(vinyl alcohol) composite attained a thermal conductivity of ~6.90 W/mK at a low polymer concentration of 6 wt% [52]. In another study reported by Hu et al. [53], a natural nacre-alike, morphological structure in flexible thin films of cellulose nano-fiber (CNF) was prepared by sonication and vacuum-assisted filtration method. During the process, h-boron nitride was initially hydrated and then exfoliated before the filtration process. Hence, the results showed that with only a 25 wt% loading concentration of hydroxylated boron nitride nano-sheets (BNNS-OH), the in-plane tensile strength and thermal conductivity of BNNS-OH/CNF thin films were measured to be approximately 50 MPa and 22.67W/mK. Contradictory to the hot-pressing technique, the vacuum-assisted filtration process can evade the high-temperature treatment, which is quite adequate for temperature-sensitive polymer composites (Fig. 10).

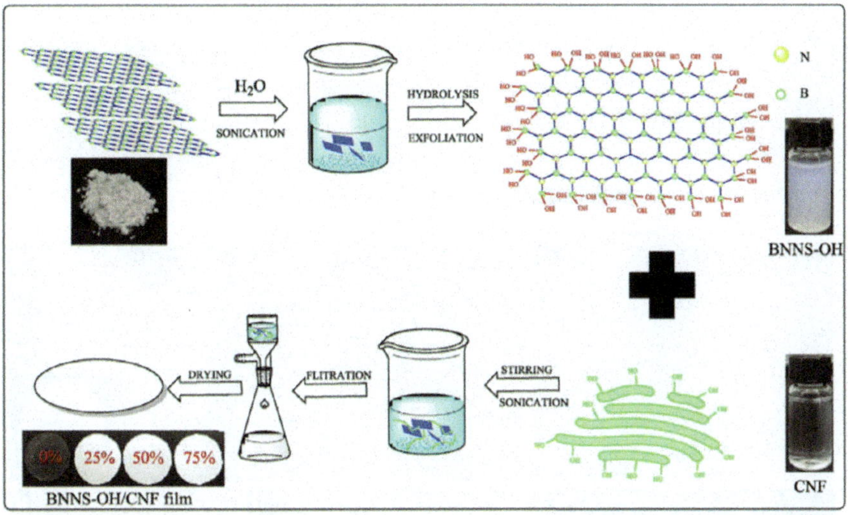

Fig. 10 Preparation and synthesis of BNNS-OH/CNF films. Reprinted with permission from Hu et al. [53]. Copyright (2018) Elsevier

3.3 Template Assembly

The template assembly technique is a common method with a unique approach to form a 3D filler-based network in the polymer system showing a substantial capability to increase the thermal conductivity with a minor filler loading concentration. The 3D-based conductive networks are usually pre-assembled and then filled with polymers by certain reported methods. The ice-template method is one of the templates. The assembly of an ice template generally starts with preparing an aqueous suspension with adhesive and the fillers, followed by the anisotropic freezing of the aqueous suspension to drive the fillers into the space between the ice template for the formation of conductive pathways. Subsequently, the ice template and the organic molecules are allowed to be removed in the freeze-drying process, leaving behind the original free-standing structure of 3D-oriented fillers. To illustrate this method, polymer composites were prepared after infiltration of the epoxy resin into the pre-constructed 3D fillers' network. By this technique, Zeng and co-workers have achieved a high in-plane thermal conductivity of around 2.85 W/mK at a low boron nitride loading concentration of around 9.29 vol% via ice-template assembly [54] (Fig. 11).

The template assembly method is often used for polymer composites. A robust support material (often with good thermal and mechanical properties) serves as the template for the filler to adhere to, resulting in enhanced isotropic thermal conductivity as a continuous 3D network of filler material is achieved. For example, Hu et al. enhanced the thermal conductivity of 3D boron nitride network-based epoxy resin composites up to 4.42 W/mK at a 34 vol%, which is quite higher than 1.81 W/mK from 3D-boron nitride-based epoxy composites [55], shown clearly in Fig. 12.

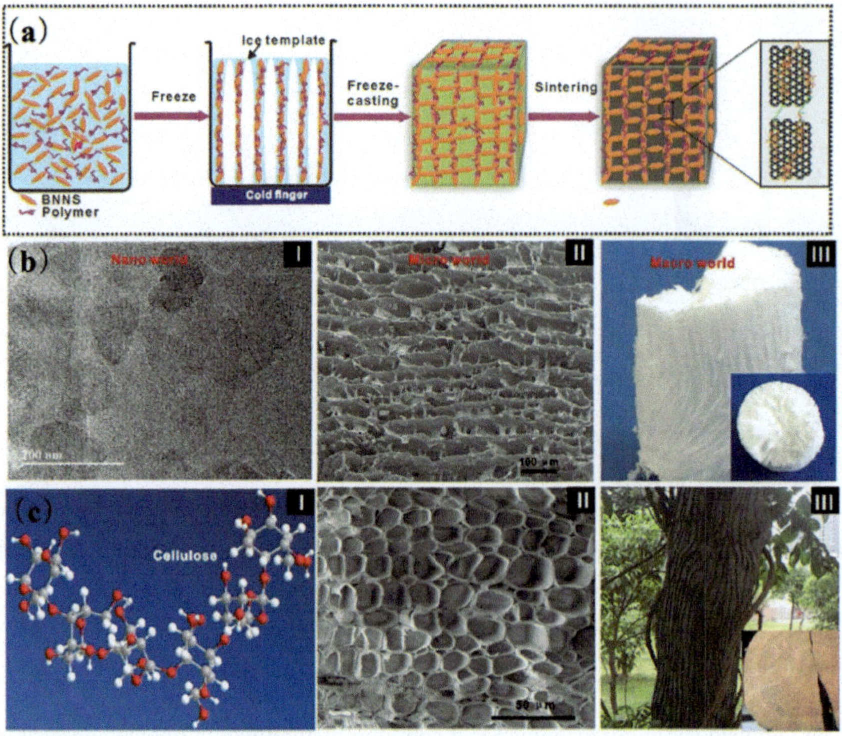

Fig. 11 Schematic representation of the template assembly process. Reprinted with permission from Zeng et al. [54]. Copyright (2015) WILEY-VCH Verlag GmbH & Co. KGaA, Weinheim

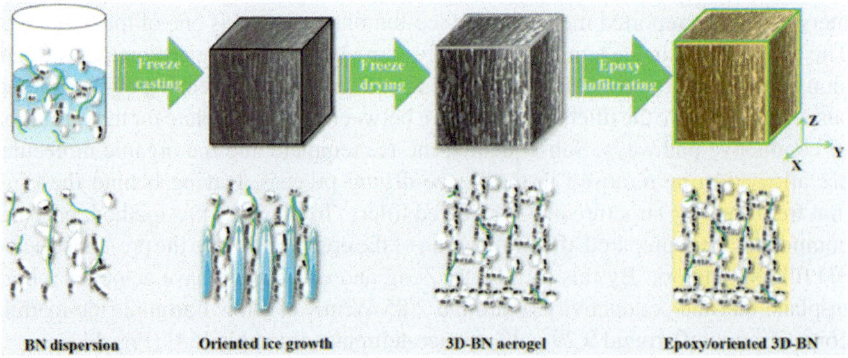

Fig. 12 The construction process of ice template assisted ordered 3D network [30]. Reprinted with permission from Tsakalakos L [30]. Copyright (2010) Taylor and Francis

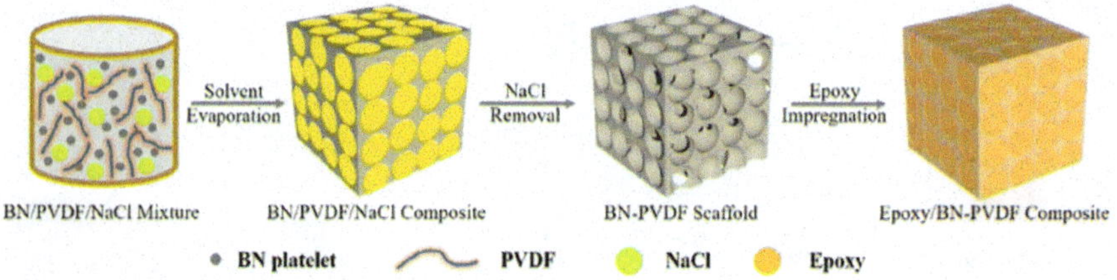

Fig. 13 Schematic diagram of the synthesis procedure for boron nitride-polyvinylidene fluoride scaffolds and epoxy/boron nitride-polyvinylidene fluoride composites. Reprinted from permissions from Chen et al. [56]. Copyright 2020 American Chemical Society

In addition to using ice as a template, some additional materials have also been used to create 3D networks. Chen et al. [56] used salt as a template material to mix boron nitride and polyvinylidene fluoride in acetone. Since the salt isn't soluble in acetone, a stable network is constructed by moving boron nitride into the region of salt particles after removing salt and the evaporation of acetone, as shown in Fig. 13. After this step, it was followed by the steps including vacuum-assisted impregnation of epoxy by penetrating the vacancies among the boron nitride polyvinylidene fluoride scaffold. The epoxy-based composites were formed with a conductive boron nitride-polyvinylidene fluoride network achieving a thermal conductivity of 1.23 W/mK with a filler fraction of 21wt% boron nitride. The polyvinylidene fluoride employed was used to increase the viscosity and inhibit the sedimentation of the boron nitride particles. Subsequently, thermal treatment of the epoxy-based boron nitride-polyvinylidene fluoride composites was further converted into epoxy-based boron nitride-carbon composites. As a result, the thermal conductivity of composites rose to 1.47 W/(m·K) due to the mitigation of the phonon scattering phenomenon at the interfaces.

3.4 3D Printing

Three-dimensional (3-D) printing technique has been utilized extensively in manufacturing and research institutes, labs, and industries around the globe. Gurijala et al. [57] and co-workers have combined the alignment of the magnetic field, vibration, and stereo-lithographic 3D printing leading to a state-of-the-art manufacturing platform for fabricating printable dielectric composites (Fig. 14). The h-boron nitride-based polymer composites achieved a high thermal conductivity of around 9.0 W/mK with a high 60 vol% filler content loading. This incredible augmentation of thermal conductivity is due mainly to the thermal conductive pathways originating from 3-D printed-based magnetic alignment and percolation. This research work has shown a programmable scaffold to print the thermally conductive polymer composites without restraint, guiding the heat dissipation channel for the electronic devices intelligently.

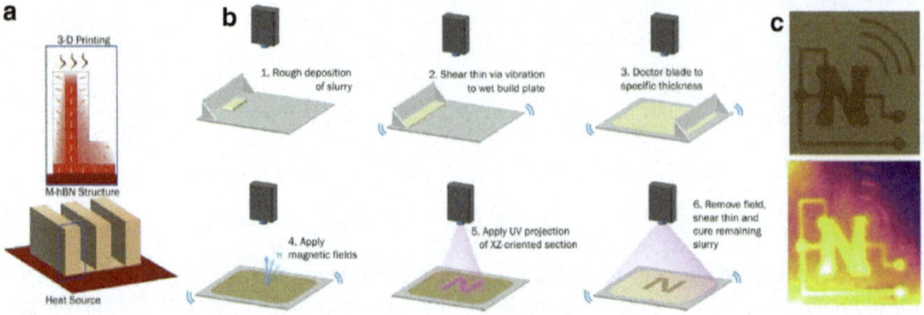

Fig. 14 Schematic drawing of 3D magnetic printing process. Reprinted with permission from Gurijala et al. [57]. Copyright (2020) Elsevier

3.5 *Melt Impregnation*

The melt-impregnation technique is considered the most facile, intuitive, and standard method for preparing composite material. It typically involves melting the solid powder or the filler's phase change material via blending mixing techniques like probe-sonication/ultra-sonication, and magnetic or mechanical stirring at high rpm, after quenching/crystallization at room temperature. The technique is excellent for the nanofillers' uniform dispersion, which is pertinent to achieving the percolation threshold for enhanced composite characteristics.

To illustrate this mixing procedure, Amine et al. [58] prepared beeswax/graphene-based phase change materials (PCM) composite using a blend mixing technique using ultra-sonication to assure uniform homogeneous filler distribution for a boosted thermal conductivity of composite material. This technique is also usually used to infiltrate porous fillers, as explained by Chen et al. [59] to prepare PCM composite material by employing a 3D carbon fibers matrix (3D Cf/SiC-ZrC-ZrB_2). Likewise, Karthik et al. [60] also reported the preparation of erythritol–graphite foam-based PCM composite using a similar method.

3.6 *Vacuum Impregnation*

The vacuum impregnation method involves impregnating a matrix material, mostly polymers like PCM, paraffin, etc., into the fillers involving mostly clay nano-platelets, graphene structures, graphene nanoplatelets, etc. They are initially exposed to the vacuum to remove any air bubbles present within the filler material's pores, vacancies, crevices, and cracks [61–63]. Once adequate degassing has been achieved, the liquid phase-matrix material, i.e., PCM, is added into the system under a vacuum at a high temperature to preserve the liquid phase at a much lower viscosity [61]. Some researchers also demonstrated their research by using solvating agents like acetone or toluene in composites, considering their compatibility with the organic or inorganic

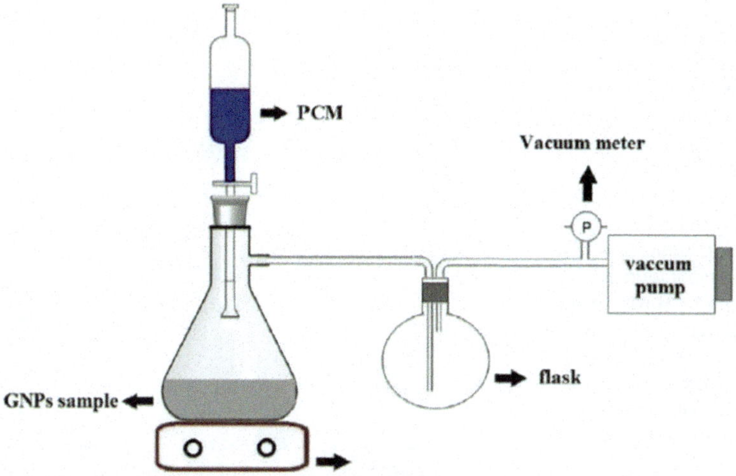

Fig. 15 Typical experimental apparatus setup for the vacuum impregnation process. Reprinted with permission from Mehrali et al. [61]. Copyright (2013) Elsevier

PCM [61, 64]. The addition of these solvents lowers the viscosity of the PCM material to permit an effective and rapid infiltration of the pores, followed by heat treatment for the evaporation of solvents [61]. To explain this technique, the experimental setup of the system is portrayed in Fig. 15.

3.7 Solution Impregnation

Also referred to as the incipient wetness impregnation [60] or capillary impregnation method [65] is a commonly employed technique for a range of applications, such as the catalyst synthesis [66, 67], deposition of nanoparticles [68], and preparation of nanocomposites [69]. This method is a highly effective approach for impregnating filler materials into a given matrix material. The procedure is relatively simple to set up, install, and use. A graphical illustration of the method is outlined in Fig. 16 [70].

The filler material is first dispersed in a solvent, followed by the immersion of the matrix material in the solution. Additional homogeneous dispersion can be attained through ultrasonic probe/bath agitation; nonetheless, it differs depending on the compatibility and the nature of the matrix and filler material. When uniform dispersion is attained, the base solvent, usually with a low boiling point, is evaporated by heating the remnants of the matrix and filler [60, 64, 70–72].

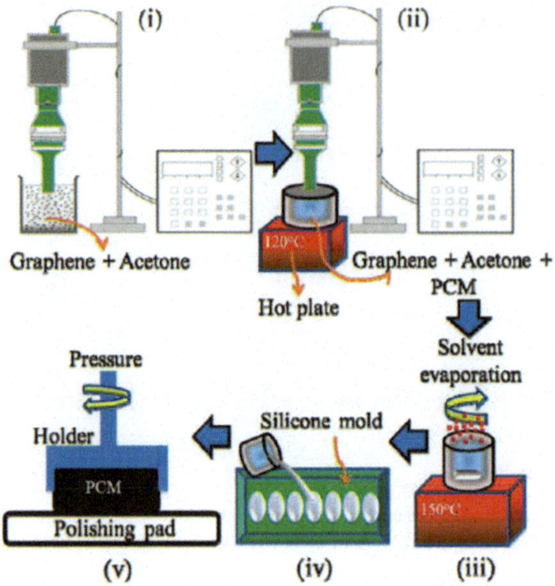

Fig. 16 Schematic Schematics of the fabrication process of PCM-based graphene composites: (i) dispersion and mixing of graphene in acetone via ultrasonic agitation, (ii) dispersion and mixing of graphene and PCM in acetone via sonication, (iii) solvent evaporation using a hot, magnetic stirring plate, (iv) Placement of the molten PCM in the mold, and (v) polishing the specimen. Reprinted with permission from Yavari et al. [70]. Copyright (2011) Elsevier

4 Composite Amelioration via Composite Properties

4.1 Adjusting Pore Morphology

The thermal conductivities of final assemblies are affected by the unique pore morphologies of 3D-based h-BN structures, which are created by different assembly and fabrication techniques. 3D h-BN can be grown on 3D metal templates, for instance, metal foams and meshes. The individual cellular structures of Ni-based foams are mimicked by Boron nitride nanofibres (BNFs) grown on Ni foam templates [73]. Large open pores, with sizes ranging from 200 to 500 μm, are revealed by the triangular-shaped hollow struts, which are made up of thin, continuous, and polycrystalline boron nitride cell walls. Due to BNFs' high porosities and thin walls, their densities can be as low as 1.3 mg cm^{-3}, resulting in much lower thermal conductivity for hollow foams compared to solid struts. This finding in the literature suggests that gas conduction influences the thermal conductivities of porous BNFs via large open and closed pores. In contrast, solid conduction is mainly devoted to the strut walls, which is almost negligible [74, 75]. Here, in aerogels, thermal conduction contributions through the gas/air and solid are typically referred to as 'gas conduction' and 'solid conduction', respectively [76].

The inter-connected, porous-based 3D networks can be used as 3D fillers for thermally conductive composites due to their high thermal conductivities of struts made up of BN sheets. Homogeneous, continuous, and interconnected BN sheets can be easily fabricated into composites through the permeation of liquid polymers into the open pores surrounding the cellular struts [73]. The thermal conductivities of these composites are significantly improved at low filler contents, mainly due to the unimpeded thermally conductive paths created by the continuous, interconnected 3D networks of BN. The porous structure can be further modified by using templates with contrasting morphologies. The cell wall thickness and crystallinity of nanosheets can be optimized by regulating CVD parameters such as the composition of precursor gases, the annealing time of the template, and the feeding time of gas [77–79].

Another approach to constructing 3D networks of BN sheets is through a self-assembly process using a sol–gel method. However, BNNS cannot form LCs in the solution because of their low aspect ratios, so binders must be added to facilitate gelation and maintain the structural integrity of aerogels after the freeze-drying technique [80, 81]. A cellulose-BNNS composite aerogel is produced by using cellulose fibers as a crosslinking agent in the dispersion of boron nitride, resulting in interconnected BNNS cells that are bound by cellulose fibers [80]. BNNS hybrid aerogels typically have randomly distributed pores, but pores that are aligned in 3D structures are generally more useful for thermal management applications. For example, when BNNS are added to the aligned cell walls of freeze-cast aerogels, their high in-plane thermal conductivities are fully utilized, resulting in excellent heat dissipation characteristics in the direction of the aligned fillers, making them suitable for thermal interface materials (TIMs) [82]. The unidirectional freeze-casting method involves creating a BNNS structure using ice crystals, forming a 3D structure that has BNNS cell walls and pores oriented in the direction of ice growth [83]. The growth of ice crystals is directed by the temperature gradient created in the dispersion between the cold metal bottom and the room-temperature top. The pore morphologies can be tailored by adjusting freezing conditions, for instance, aligned pores with random cell walls are obtained using a single temperature gradient, while highly aligned cell walls are achieved by utilizing dual temperature gradients.

The thermal conductivities of 3D designs and composites are also determined by the structures and compositions of cell walls in addition to pore morphologies. Cell walls that have high conductive fillers with low interfacial thermal resistance are ideally required [84–86]. An effective approach is to incorporate thermally conductive boron nitride nanosheets (BNNS) layers into micro-sandwich polyimide (PI) composites by using a sequential bidirectional freeze-casting method, with co-aligned layers [87]. The aligned cell walls featuring alternating rGO/BNNS/rGO layers, result in lower interfacial resistance between the fillers compared to mixed fillers. This leads to a significantly higher thermal conductivity after the composite is transformed into a thin film through the compaction process.

BN aerogels that have thin cell walls consisting of nanosized pores are associated with thermal insulation [76]. They are made by CVD on GA-based templates using borazine as a precursor. These aerogels have double pane BN cell walls of around 1–100 nm in thickness depending on the concentration of precursors and nano-sized pores between the two panes formed after etching out of GA templates [76]. The thin walls decrease solid conduction while the nanosized pores lessen gas conduction, resulting in an ultra-low thermal conductivity of only 20 $mWm^{-1}\ K^{-1}$ at a density of 10 mg cm^{-3}.

4.2 Adjusting Density and Pore Size

The thermal conductivities of specimens are closely linked to the density and pore size of 3D structures. When 3D systems have high densities, it leads to small pore sizes, which in turn results in high filler loading concentrations when polymers are infiltrated. This results in high thermal conductivity of composites. For example, the template-directed CVD method is used to achieve high densities of BN 3D structures by using high-density templates [88].

An example of this is when BNFs are grown on high-density carbon monolith templates using B_2O_3 as the precursor material, [88]. The resulting BNFs replicate the carbon templates, featuring small pore sizes of less than 1 μm and a high density of 561 mg cm^{-3}, which is more than 50 times higher than BNFs grown on Ni substrate templates (~10 mg cm^{-3}) [80]. Composites are fabricated by infiltrating the BNFs with polymethyl methacrylate (PMMA) to achieve a high BN loading content of 25 vol %. At this high filler loading concentration, the homogenously dispersed BN network leads to a high thermal conductivity of 11.13 Wm^{-1} K $^{-1}$ for the PMMA/BNF composite. To increase the density of 3D structures in solution-based processing systems, a strategy is employed to enhance the concentration of fillers in the solution [89]. Increasing the filler concentration to 250 mg ml^{-1} in a BN-SiC hybrid composite solution results in shortened gaps between the aligned cell walls after unidirectional freeze casting and freeze-drying, leading to an increased density of BN-SiC based aerogel. After infiltrating with PDMS, the high-density BN-SiC scaffolds ultimately result in higher filler loading contents of over 8 vol %. While high densities are necessary for thermal heat dissipation applications, low densities are preferred for thermal insulation applications of 3D structures. As such, low filler concentration solutions are used to prepare aerogels with ultra-low densities.

4.3 Structural Engineering of the Cell Wall

The thermal properties of BN structures can also be influenced by the detailed nanoscopic structures of the cell walls, which can vary the intrinsic thermal conductivities of BN sheets. The high intrinsic thermal conductive properties of BN particles stem from their pure crystalline structures, which enable the uninterrupted transport of phonons without scattering [90]. Therefore, methods that can preserve the crystallinity of BN are sought after to maintain their high thermal conductivities. CVD is one of the most effective techniques for creating 3D BN structures with high thermal conductivity in cell walls. Freestanding BN flakes made by CVD have porosities of over 99%, with thermal conductivities as high as 0.84 Wm^{-1} K^{-1}. This increase is mainly due to the high intrinsic thermal conductivities of BN cell walls [91], indicating that high porosity may not be sufficient to produce low thermal conductivity for thermal insulation applications. The thermal conductivities of cell walls should also be suppressed as much as possible. One of the compelling ways is to use functionalized fillers as a substitute for pristine ones.

In summary, thin surface-modified 2D nanosheets with many interfaces in the cell walls are crucial for achieving low thermal conductivities in 3D structures. On the other hand, thick, pure BN layers are preferred for 3D structured fillers in highly thermally conductive composites.

5 Performance Assessment Metrics for PCMs

While the scientific community has, without a doubt, made leaps of progress in designing novel composites with a variety of unique properties that can be extremely useful for several thermal management applications. However, tailoring an 'ideal' composite that fulfills all the star qualities for any application is quite unlikely. One may get close to the ideal; however, realistically speaking, there is always a compromise in some aspects of this novel composite material. For example, it is fairly simple to achieve a highly enhanced thermal conductivity at high loadings of thermally conductive filler material, but at what expense? The high filler loading translates into a low PCM content, thereby reducing the composite's energy storage capacity. So, the 'true' performance of the material in a practical scenario may not live up to the expectations.

With the advent of a variety of 1D-, 2D-, and 3D nanofillers with outstanding properties and specific synthesis techniques, as discussed earlier, each new composite has something unique to offer. Combine this variety with another massive array of matrix materials, and we have an infinitely large possibility of composite materials. Some composites may boast low filler loadings, and some might offer higher thermal conductivity, while others may offer form-stability or other desirable traits—making the task of choosing a suitable PCM composite for an application an immensely complicated one.

Traditionally, researchers have resorted to using the intrinsic properties of PCM composites as the principal figure of merit. Some popular intrinsic properties often deliberated in literature are thermal conductivity (κ_c), melting onset temperature (T_m), and the specific latent heat of melting (ΔH_m). κ_c determines how rapidly the composite absorbs or releases thermal energy or can be considered a measure of its resilience to thermal shocks. Similarly, T_m governs the suitability of the material for a particular application given the temperature constraints of the application, while ΔH_m dictates the energy capacity of the material per unit mass or volume. Notwithstanding the convenience of using these to filter out composites for a particular application, they cannot be used as a figure of merit for their actual performance in these applications. The performance of these composites in a thermal management system (TMS) is again not a simple function with just one parameter—it is a combination of several key parameters. These key parameters can be; the type of PCM, its latent heat, melting onset temperature, the filler type, composition, viscosity, and stability, to name a few.

Instead, another possible metric is the 'cooling capacity' (q) of a TMS, which is analogous to the instantaneous heat transfer from the surrounding environment into the thermal energy storage material (or PCM). A large cooling capacity is vital for a resilient TMS: one that can swiftly respond to thermal shocks and adequately buffer thermal spikes on short timescales, avoiding large temperature gradients in the TMS. However, this cooling capacity is not an intrinsic property. Instead, it depends on the TMS devices' geometry and scale, making its use ineffective as a universal performance metric. Although some may be tempted to use thermal conductivity, κ_c, as a proxy for the same cooling capacity, it disregards the impact of other intrinsic material properties described earlier (e.g., κ_c, T_m, and ΔH_m) in the heat transfer process. Consequently, a figure of merit (FOMq) that captures the role and dependencies of the intrinsic materials properties on the instantaneous rate of specific thermal energy transfer would greatly facilitate material comparison and selection for a particular TMS application.

To this end, Shamberger [92] proposed reconsidering the cooling capacity metric at an arbitrarily small scale and simplifying the problem down to just the interface. The interfacial heat transfer rate, i.e., heat flux (q"), is of prime importance when considering the performance of any material in a TMS. Thus, removing the geometrical dependencies would allow for a universal comparison of materials based on their merits. It can also help distinguish materials that otherwise boast similar properties but are essentially different, e.g., two composites may have the same thermal conductivity but have different filler or matrix materials and consequently different intrinsic properties. The following section describes the derivation of the FOMq as a function of thermophysical properties considering the analytical solution to the melting of a solid semi-infinite plane under constant-temperature boundary conditions.

5.1 Neumann-Stefan Problem: A Brief History

Phase-change problems cannot be considered as relatively new problems. One can come across several common examples in nature, for example, the melting and freezing of ice caps or the solidification and convection of molten lava and its influence on tectonic activity. This phenomenon and its associated technology have been with humankind since the Neolithic. Moreso, the evolution of our civilization, especially our current era of post-industrial development, is shaped not by the type of cast metal implements or weapons as was the case for the Copper/Bronze/Iron ages. Instead, it is strongly governed by the level of understanding we have of the solid–liquid phase change. Almost all aspects of pure sciences like astrophysics, meteorology, and geophysics have had to understand this phase-change phenomenon to advance the field. Modeling the global weather, studying the planet's core solidification, desalination of seawater, ground freezing for construction, and concrete hardening are just a few examples of the multiple encounters with phase-change problems.

In the mid-1700s, Joseph Black, a physician-cum-chemist by profession, introduced the concept of latent heat to describe the otherwise inexplicable nature of solid–liquid phase change. The process otherwise could not simply be explained under the framework of sensible heat alone. Subsequently, other great names, including Jean Baptiste Joseph Fourier, Gabriel Lame, Emile Clapeyeron, Franz Neumann, and Jozef Stefan, developed the physics and mathematics necessary to solve heat transfer problems analytically. However, Stefan's contributions to the solid–liquid phase change problem are of paramount importance since they established the foundations of the phenomenon, especially moving boundary problems. The moving boundary problems are often referred to as the "Neumann-Stefan" problem for Stefan's [93] early work on ice formation in the Arctic seas and Neumann's [94] analytical solution to the case where the boundary is maintained at a constant temperature.

5.2 FOMq Cooling Capacity

Neumann-Stefan problems typically involve a dynamically changing boundary that the governing equations of the system can define. The exact solution of these systems can be obtained for relatively simple geometries and a few special boundary conditions. In short, one can calculate the magnitude of the energy flux across the interfaces in the system along with the solidification front, i.e., the location of the solid–liquid interface along a predefined axis, using these solutions. Many have considered subtle variations to these solutions, and more details can be found in Refs. [95, 96].

Shamberger's adaptation of the Neumann-Stefan problem for assessing the performance of a PCM in a TMS device allows for the development of a figure of merit (FOMq) for the cooling capacity as a function of the intrinsic material properties. Considering the two cases, i.e., the one-region and two-region Neumann-Stefan problem, Shamberger develops the analytical solution for both. The one-region case considers a semi-infinite plane of a solid PCM at its melting temperature responding to a thermal shock at the container wall, i.e., analogous to a rapid rise of temperature at the container wall, while the solid–liquid interface is at an arbitrary position $x = \delta(t)$ at $t > 0$. For the more general, two-region case, the solid phase is considered to be at $T = T_o < T_m$ and both phases are considered active regions. Solving for the flux $\left(q''(x,t)\right)$ at the interface, where x = 0, for both the cases gives almost similar solutions, differing only due to the material-dependent parameter λ. This parameter is obtained by solving the transcendental Eq. 8 (Table 1). The analytical solutions of both cases are summarized in Table 1.

Equation 11 for the one-region problem and Eq. 19 for the two-region case both give the heat flux at the interface of the PCM material as a function of time. As described earlier, the function of a PCM in a TMS is to absorb thermal shocks or rapidly transfer thermal energy for storage purposes. For this purpose, Eqs. 11 and 19 can estimate this heat flux for a given set of conditions taking into account the material properties of each of these materials in the process. More specifically, isolating the material-specific component of the equation, i.e., k_l, λ_2, and α_l, one

Table 1 Analytical solutions to the One-Region and Two-Region Neumann–Stefan Problem

One-Region Neumann-Stefan Problem
Initial boundary conditions:
$T_l(x = 0, t > 0) = T_w (1)$
$T_s(x = \delta(t), t > 0) = T_m (2)$
Where $\delta(t)$ is the distance of the solid–liquid interface from the boundary wall
Energy balance for the heat transfer within the liquid, while convective heat transfer in liquid is ignored, can be written as follows:
$(\rho C_p)_l \frac{\partial T_l}{\partial t} = k_l \frac{\partial^2 T_l}{\partial x^2} (3)$
Where ρ, C_p, and k_l are the density, specific heat, and thermal conductivity of the liquid phase (denoted by the subscript l)
Applying conservation of energy and temperature continuity at the solid–liquid interface
$T_s(x = \delta(t), t > 0) = T_l(x = \delta(t), t > 0) = T_m (4)$
$\rho_l L_w \frac{d\delta(t)}{dt} = k_l \frac{dT_l(\delta, t)}{dx} (5)$
Where L_W is the specific latent heat of fusion of the PCM material
The above equations conform to the classical one-region Neumann-Stefan problem for which the analytical solution is given by:
$\delta(t) = 2\lambda\sqrt{\alpha_l t} (6)$
$\frac{T_l(x,t) - T_W}{T_m - T_W} = \frac{\text{erf}(x/2\sqrt{\alpha_l t})}{\text{erf}(\lambda)} (7)$
Where α is the thermal diffusivity, and the parameter λ is obtained by solving the transcendental equation below:
$\lambda e^{\lambda^2} \text{erf}(\lambda) = St_l / \sqrt{\pi} (8)$
Where the St_l is the Stefan number for the liquid phase
$St_l = \frac{C_{p,l}(T_W - T_m)}{L_W} (9)$
The heat flux through the liquid as a function of distance (x) from the source wall and time is given by:
$q"(x, t) = -k_l \frac{\partial T}{\partial x} (10)$
Solving for the heat flux at the melting front and the source wall, we obtain the following:
$q"(0, t) = \frac{T_W - T_m}{\sqrt{t}} \cdot \frac{1}{\sqrt{\pi}\text{erf}(\lambda)} \cdot \frac{k_l}{\sqrt{\alpha_l}} (11)$
$q"(\delta(t), t) = \frac{T_W - T_m}{\sqrt{t}} \cdot \frac{1}{\sqrt{\pi}\text{erf}(\lambda) e^{\lambda^2}} \cdot \frac{k_l}{\sqrt{\alpha_l}} = \frac{\rho_l L \lambda \sqrt{\alpha_l}}{\sqrt{t}} (12)$
$q(0, t) = q(\delta(t), t) . e^{\lambda^2} (13)$
Two-Region Neumann-Stefan Problem
For the case with two active regions where the solid phase is at an initial temperature well below the melting temperature
$T_s(x = \delta(t), t = 0) = T_0 < T_m (14)$
Temperature profiles for the liquid and solid regions are defined as:
$\frac{T_l(x,t) - T_W}{T_m - T_W} = \frac{\text{erf}(x/2\sqrt{\alpha_l t})}{\text{erf}(\lambda_2)} (15)$

(continued)

Table 1 (continued)

One-Region Neumann-Stefan Problem
$\frac{T_s(x,t)-T_0}{T_m-T_0} = \frac{\mathrm{erfc}(x/2\sqrt{\alpha_s t})}{\mathrm{erfc}(\lambda_2\sqrt{\alpha_l/\alpha_s})}$ (16)
Where the subscript s denotes the solid phase and the parameter λ_2 is obtained by solving the transcendental equation for the two-phase case:
$\lambda_2\sqrt{\pi} = \frac{St_l}{\exp(\lambda_2^2)\mathrm{erf}(\lambda_2)} - \frac{St_s\sqrt{\alpha_s}}{\sqrt{\alpha_l}\exp(\lambda_2^2\alpha_l/\alpha_s)\mathrm{erfc}(\lambda_2\sqrt{\frac{\alpha_l}{\alpha_s}})}$ (17)
Where St_s is the Stefan number for the solid phase
$St_s = C_{p,s}(T_m - T_s)/L_W$ (18)
The heat flux across the boundary wall as a function of time, for the two-region case, is given by:
$q''(0,t) = \frac{T_W - T_m}{\sqrt{t}} \cdot \frac{1}{\sqrt{\pi}\mathrm{erf}(\lambda_2)} \cdot \frac{k_l}{\sqrt{\alpha_l}}$ (19)

can readily determine the influence of these properties on the interfacial flux. Thus allowing the comparison of two different materials based on their relative merits. From Eq. 19, Shamberger proposed the following materials-based figure of merit:

$$FOM_q = \frac{k_l}{\sqrt{\alpha_l}\mathrm{erf}(\lambda_2)} = \frac{\sqrt{k_l\rho_l C_{p,l}}}{\mathrm{erf}(\lambda_2)} \quad (20)$$

Subject to the same external parameters, such as system geometry and magnitude of thermal shock, a material with a larger FOMq would absorb the thermal energy at a much quicker rate and is thus more capable of regulating thermal spikes in shorter periods. FOMq value is directly proportional to the interfacial heat flux at the PCM boundary, within the framework of assumptions made earlier (see Table 1). Also, note that the FOMq is not entirely independent of the extrinsic factors (like subcooling temperatures) as these are implicitly incorporated in the parameter λ_2 which depends upon St_s and St_l, the Stefan numbers for the solid and liquid phases, respectively. Therefore, the specific heat in the two-region problem is also explicitly included in this FOMq metric.

The case considered above is for a melting solid, where the heat at the container wall is transported through the liquid to the melt front, at δ(t) from the container boundary or wall, where the PCM absorbs it as it melts further. For this reason, the FOMq is characterized by the properties of the liquid phase. Alternatively, the solidifying case can also be considered, given the symmetrical nature of both problems. The FOMq can then be redefined for the solidification problem, i.e., heat transfer from the PCM to the environment. The FOMq for this solidification process would have the same characteristic material parameters but now for the solid phase instead of the liquid phase. Depending upon the nature of specific applications, these metrics can prove to be a more reliable and relevant metric for critically comparing the performance of different materials for the same application. In general, to compare the FOMq of different materials, one must first establish the environmental conditions to estimate the Stefan numbers for the liquid and solid phases. Subsequently, the parameter λ_2 is evaluated by solving the transcendental Eq. 15, after which the evaluation of the FOMq is pretty straightforward.

5.3 *Improving FOMq*

This figure-of-merit approach simplifies the performance evaluation of a particular PCM material and is helpful in terms of practical application. However, working with composite materials brings another challenge—telling apart two composites with the same FOMq but entirely different filler materials. There are instances where one can achieve more or less the same FOMq numbers; however, the materials can be entirely different, i.e., one may have a high filler loading (>20 wt%) for a particular matrix material; in contrast, the other may have a low filler loading (<10 wt%) but with an entirely different matrix material. So, what to do now?

To address this problem, one can simply normalize the FOMq with the filler loading to elucidate the contribution based on the filler. It is particularly useful when deciding between composites where the cost is a factor. In general, it would be preferable to use a material that requires the minimum amount of filler to avoid unnecessary material costs. The filler loading normalized FOMq (FOMq') can also help distinguish between form-stable PCM composites—composites that do not lose their form or shape even past the melting temperature of the matrix material. These composites are often characterized by high filler loadings (often >50 wt%).

In addition to the above normalization, it is also possible to consider other material aspects like the latent heat or the Young's modulus. A combination of the two can also prove helpful depending upon the nature of the decision involved. In some cases, thermal performance alone is not critical, as is the case for TIMs. The total thermal resistance of a TIM, often the selection criteria, is a combination of the contact resistance and the bulk conductivity of the TIM [97]. Two materials with the same thermal properties or FOMq may satisfy the performance criteria; however, for TIMs, the added requirement is its' capacity to effectively 'wet' the two interfacial surfaces is of paramount importance. Suppose the TIM does not perfectly adhere to the surface. In that case, the resulting air cavities compromise the integrity of the contact and form an insulating barrier for the thermal pathway resulting in an unnecessary loss of performance. TIMs are, in general, designed to conform well to surface irregularities to help minimize air pockets, mechanical stress on components, and enhance thermal conductivity across the interface [97]. Therefore, the thermo-mechanical properties also need due consideration in the design process. For example, the modulus and stress relaxation of the material at the typical operating temperatures of the TMS device must be carefully taken into account.

5.4 Performance Assessment: FOM_q Analysis

To highlight the significance of the FOMq analysis, we briefly compare different composite materials and the implications of this figure-of-merit approach in this section. For a detailed analysis, readers are directed to reference [11]. While PCM-based nanocomposites are gaining popularity, focus on BN is still limited, and thus only a limited number of studies were available for the current analysis.

To gain insights into the progress of PCM composites with BN or carbon-based nanofillers, we extracted the material properties of 82 composite materials from published results and calculated the FOMq metric for each of these materials. For properties that were not reported in those studies, appropriate estimates were used instead. The data is summarized in Tables 2 and 3, and a graphical summary is presented in Fig. 17. With this available data at our disposal, we investigated whether factors other than the filler type or composition significantly affected the composite performance, aka FOMq.

To this effect, we represented the data in the form of a boxplot (see Fig. 17) where FOMq and FOMq' (normalized to filler wt%) were plotted against the method used for synthesis. A noticeable trend was observable between the synthesis route and the FOMq/FOMq' for both types of nanofillers (carbon-based and BN). Several interesting observations can be made about these composites from the data extracted, especially when considering the type of nanofiller employed. For this analysis, carbon nanotubes (CNT) and carbon nanofibers (CNF) were grouped into CNT. The results are summarized in Fig. 18.

While at first look at the FOMq values, one may assume that CNTs perform poorly in terms of enhancement; however, the FOMq' values reveal that they outperform other nanofillers. There are two possible reasons for this. Firstly, the aspect ratio of CNTs sets them apart from the other fillers, which are some form of platelets. Secondly, their large aspect ratio possibly helps them reach the percolation threshold at much lower filler loadings than other composites.

The percolation threshold, the minimum amount of a filler required to establish long-range connectivity in a random system, is critical to enhancing the thermal conductivity of a composite material. For a typical platelet-type geometry with a low aspect ratio, the said percolation threshold is attained at much larger filler loadings due to the geometrical constraints. CNTs can have aspect ratios as high as 2000, whereas the platelet geometry can rarely match such a high number [95]. In a separate study [116], we investigated several hybrid combinations of different nanofillers. We observed that introducing a nanofiller with a high aspect ratio (like CNTs or CNFs) significantly improved the thermal conductivity of the composite. An illustration of the said phenomenon is provided in Fig. 19. Such nanofillers' elongated and tortuous physique helps provide an interconnected spaced network with improved thermal conductivity and diffusivity within the composite. Thus, apart from helping evaluate composite performances, using such a figure of merit can also prove helpful in making strategies for novel composite materials.

Table 2 Summary of past works based on boron nitride nanofillers. Reprinted with permission from Ref. [11]

Synthesis route	Matrix	Filler	Filler [wt%]	$\kappa\left[\frac{W}{m.K}\right]$	T_m [°C]	ΔH_m $\left[\frac{J}{g}\right]$	$\frac{\kappa_C-\kappa_O}{\kappa_O}$ [%]	FOM_q	FOM_q'	Ref.
Melt Mixing	Paraffin	h-BN nanosheets	10	0.53	59	177	165	4108	411	Fang et al. [98]
	Stearic Acid - Octadecane Eutectic	h-BN	10	0.32	28	208	8	3209	321	Su et al. [99]
	Paraffin	h-BN	26	0.50	52	135	178	3889	150	Qian et al. [44]
	Eutectic Hydrate Salts	BN Foam	4	1.36	27	213	937	8607	2152	Han et al. [43]
Solution Impregnation	PEG SB-20	h-BN	20	1.24	57	142	288	5864	293	Yang et al. [100]
	PEG	h-BN	30	0.79	63	124	155	4649	155	Yang et al. [101]
	PEG	h-BN/GN (30 / 1)	31	1.33	64	122	329	6026	194	Yang et al. [101]
	PVA	h-BN	27	1.63	228	159	443	7326	271	Zhang et al. [102]
	PEG	h-BN	30	2.77	62	116	823	8444	281	Yang et al. [103]
	PEG	h-BN/GO (30/4)	34	3.00	65	107	900	8661	255	Yang et al. [103]
	Thermoplastic Polyurethane (TPU)	h-BN	40	1.20	65	74	532	4971	124	Yu et al. [104]
	Polystyrene (PS)	h-BN/GNP (1.5:20)	21.5	0.67	240	80	330	3201	149	Cui et al. [105]
	Polystyrene (PS)	h-BN/GNP (20:1.5)	21.5	0.28	240	80	77	2057	96	Cui et al. [105]
	Polyamide 6 (PA6)	h-BN/GNP (1.5:20)	21.5	1.76	269	192	524	7859	366	Cui et al. [105]
Vacuum Impregnation	Bio-based PCM	BN PT 160	14	0.73	30	117	377	4044	283	Jeong et al. [106]
	Paraffin	h-BN nanosheets	40	3.47	42	80	1097	8815	220	Yang et al. [107]
	PEG c20	h-BN/GO	19	1.52	60	146	375	6542	344	Yang et al. [100]

(continued)

Table 2 (continued)

Synthesis route	Matrix	Filler	Filler [wt%]	$\kappa\left[\frac{W}{m.K}\right]$	T_m [°C]	ΔH_m $\left[\frac{J}{g}\right]$	$\frac{\kappa_c-\kappa_o}{\kappa_o}$ [%]	FOM_q	FOM_q'	Ref.
	PEG C20	h-BN/GO (<freeze temp)	19.2	1.84	60	146	475	7213	376	Yang et al. [100]
	Polyamide 6 (PA6)	GO	6.80	0.48	269	192	142	4148	610	[108]
	Polyamide 6 (PA6)	h-BN/GO (1.6/6.8)	8.40	0.89	269	192	355	5703	679	Shao et al. [108]
	Paraffin	h-BN scaffold	26	1.33	52	135	1008	6343	244	Qian et al. [44]
	PEG	3D-hBN-oriented	10	4.76	60	137	1344	10,615	1094	Lei et al. [109]
	PEG	BNNs-g27/CNF3	3	0.59	45	151	111	3731	138 2	Wan et al. [110]
	SA	hBN-E	72	0.45	52	136	73	4446	61	Xie et al. [111]
Melt Mixing	PEG	hBN	10	2.59	60	137	685	7830	807	Lei et al. [109]
Micro-encaps	Octadecane	BN-Melanine Shell	3	0.11	26	131	35	1505	436	Xia et al. [112]

Table 3 Summary of past works based on carbon nanofillers. Reprinted with permission from Ref. [11]

Synthesis route	Matrix	Filler	Filler $[wt\%]$	κ $\left[\frac{W}{m.K}\right]$	T_m $[^\circ C]$	ΔH_m $\left[\frac{J}{g}\right]$	$\frac{\kappa_c-\kappa_o}{\kappa_o}[\%]$	FOM_q	$FOM_q^{'}$	Ref.
Melt Mixing	PEG 10,000:TDI: NNDHEA	Carbon Black	1.96	0.28	51	101	5	2405	1227	Wang et al. [113]
	Paraffin	CNF	5	0.31	58	185	17	3039	608	Fan et al. [114]
	Palmitic Acid (PA)	CNT	1	0.33	61	184	50	3103	3103	Wang et al. [115]
	Paraffin	GNP	5	0.72	58	187	176	4752	950	Fan et al. [114]
	Paraffin	GNP	10	0.367	38.9	116	43	2823	282	Shahid et al. [116]
	Paraffin	GNP/TRG (8:2)	10	0.366	38.8	120	43	2859	286	Shahid et al. [116]
	Paraffin	GNP/CNT (8:2)	10	0.404	38.9	116	58	2979	298	Shahid et al. [116]
	Paraffin	GNP/CNF (8:2)	10	0.446	39.0	115	74	3084	308	Shahid et al. [116]
	Paraffin	L-MWCNTs	5	0.31	58	177	19	3010	602	Fan et al. [114]
	Paraffin/Diatomite/ PCM-D600	MWCNTs	0.26	1.68	27	89	42	5068	19,493	Xu and Li [117]
	Paraffin	MWCNTs	0.6	0.34	21	130	40	2684	4473	Kumaresan et al. [118]
	PEG 10,000:TDI: NNDHEA	MWCNTs	1.96	0.31	53	100	17	2518	1285	Wang et al. [113]

(continued)

Table 3 (continued)

Synthesis route	Matrix	Filler	Filler $[wt\%]$	κ $\left[\frac{W}{m.K}\right]$	T_m $[°C]$	ΔH_m $\left[\frac{J}{g}\right]$	$\frac{\kappa_c-\kappa_O}{\kappa_O}[\%]$	FOM_q	$FOM_q^{'}$	Ref.
	Paraffin	MWCNTs	2	0.32	52	164	34	2908	1454	Wang et al. [119]
	Paraffin	S-MWCNTs	5	0.34	58	178	31	3161	632	Fan et al. [114]
	PEG 10,000	SWCNTs	1.96	0.33	52	101	25	2652	1353	Wang et al. [113]
Micro-encaps (Core–Shell)	Stearic Acid + 1% SDS	Graphene	4.76	0.34	71	182	63	3223	677	Dao and Jeong [120]
	Stearic Acid + 2% SDS	Graphene	4.76	0.29	72	181	40	2969	624	Dao and Jeong [120]
Micro-encaps (Emul Polym)	Paraffin + SiO2	GO	–	1.16	50	87	78	-	-	Yuan et al. [121]
	Paraffin	SiO2	–	1.03	49	90	75	-	-	Yuan et al. [121]
Micro-encaps via In-situ Polymerization	Paraffin + Urea Formaldehyde	GO	10	0.69	32	148	208	4232	423	Qiao and Mao [122]
	n-Dodecanol	GO/ (PMF)	4	0.19	-	74	66	1534	383	Sari and Karaipekli [123]
	Paraffin + GO	UF	10	1.06	33	151	376	5152	515	Qiao and Mao [122]
Micro-encaps/ Pickering Emul	Stearic Acid	PVA@Graphene (PVA@G)	4.76	0.35	74	184	72	3250	683	Dao and Jeong [124]
Sol–Gel Method	Paraffin + SiO2	EG	7.2	0.25	28	104	95	2007	279	Li et al. [125]
	Paraffin	SiO2	20.8	0.16	28	113	28	2047	98	Li et al. [125]
Solution Impregnation	Stearic Acid	Cr-MIL-101-NH2 MOF	30	0.51	71	121	149	3765	126	Luan et al. [71]

(continued)

Table 3 (continued)

Synthesis route	Matrix	Filler	Filler $[wt\%]$	κ $\left[\frac{W}{m.K}\right]$	T_m $[°C]$	ΔH_m $\left[\frac{J}{g}\right]$	$\frac{\kappa_c-\kappa_o}{\kappa_o}[\%]$	FOM_q	$FOM_q^{'}$	Ref.
	$LiNO_3/NaNO_3$ Eutectic	EG	19	6.61	197	190	687	20,727	1091	Zhong et al. [126]
	LiNO3/ KCl Eutectic	EG	19	5.59	169	158	501	17,554	924	Zhong et al. [126]
	LiNO3/NaCl Eutectic	EG	22	4.71	228	232	489	19,273	876	Zhong et al. [126]
	Stearic Acid	Graphene	4.76	0.26	72	166	26	2700	567	Dao and Jeong [120]
	Stearyl Alcohol	Graphene	4	0.92	66	211	154	5359	1340	[70]
	Capric Acid + HNT	Graphite	5	0.76	30	75	104	3194	639	Mei et al. [127]
	Capric Acid	Halloysite Nanotube	40	0.48	29	76	29	3239	81	Mei et al. [127]
	Erythritol	Graphite Foam	25	3.77	118	251	424	16,231	649	Karthik et al. [60]
Vacuum impregnation	PEG	Cellulose-GNP Aerogel	5.3	1.35	63	156	463	6578	1241	Yang et al. [128]
	Paraffin/EP	CNT	1	0.32	36	157	45	2832	2832	Karaipekli et al. [129]
	PEG	Diatomite	50	0.32	28	87	10	2889	58	Karaman et al. [130]
	PEG-Diatomite (1:1)	EG	10	0.67	27	80	131	3481	348	Karaman et al. [130]

(continued)

Table 3 (continued)

Synthesis route	Matrix	Filler	Filler $[wt\%]$	κ $\left[\frac{W}{m.K}\right]$	T_m $[°C]$	ΔH_m $\left[\frac{J}{g}\right]$	$\frac{\kappa_c-\kappa_o}{\kappa_o}[\%]$	FOM_q	$FOM_q^{'}$	Ref.
	PA	EG	20	0.6	61	148	252	4264	213	Sari and Karaipekli [123]
	$CaCl_2.6H_2O$	EG	50	8.8	30	86	1376	16,642	333	Duan et al. [131]
	$CaCl_2.6H_2O$/5% OP-10	EG	50	4.02	28	49	574	8945	179	Duan et al. [131]
	Paraffin	EP	40	0.15	36	161	-32	2038	51	Karaipekli et al. [129]
	Palmitic Acid	GNP (300)	22	2.75	62	160	848	9435	429	Mehrali et al. [61]
	Palmitic Acid	GNP (500)	16.9	2.43	61	171	738	8918	528	Mehrali et al. [61]
	Palmitic Acid	GNP (750)	8.1	2.11	61	189	628	8307	1026	Mehrali et al. [61]
	Paraffin	Graphene Oxide (GO)	55.19	1.32	52	59	360	4747	86	Mehrali et al. [132]
	Paraffin	Nano-Graphite (NG)	10	0.94	28	182	641	5545	555	Li [133]
Solution Impregnation	SAT-SPM	GNP750	5	1.52	55	248	114	9636	1927	Mehrali et al. [134]
	SAT-SPM	GNP500	5	1.69	55	238	138	9961	1992	Mehrali et al. [134]
	SAT-SPM	GNP300	5	1.93	55	231	172	10,486	2097	Mehrali et al. [134]
Micro-encaps	Octadecane	Graphene/SiO2	2.5	1.4941	28	108	133	5021	2008	Zhu et al. [135]

(continued)

Table 3 (continued)

Synthesis route	Matrix	Filler	Filler $[wt\%]$	κ $\left[\frac{W}{m.K}\right]$	T_m $[°C]$	ΔH_m $\left[\frac{J}{g}\right]$	$\frac{\kappa_c-\kappa_o}{\kappa_o}$ [%]	FOM_q	$FOM_q^{'}$	Ref.
	Paraffin	rGO/GNP	1.58	0.9	68	187	260	4979	3151	Zhou et al. [136]
	Paraffin	rGO	1.5	0.67	69	190	168	4454	2970	Zhou et al. [136]
	SA	GO-MF	0.38	0.435	124	126	24	3058	8046	Zhao et al. [137]
	Paraffin	G-foam	0.35	0.45	52	147	75	3318	9481	Yu et al. [138]
	Paraffin	G-film-GO	1.14	11.594	52	143	4411	16,655	14,610	Yu et al. [138]

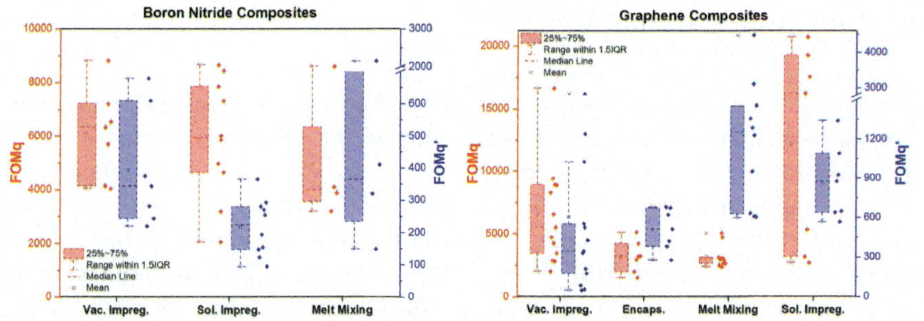

Fig. 17 Boxplot of FOM_q and $FOM_q^{'}$ for Boron nitride and carbon-based PCM composites for different synthesis approaches

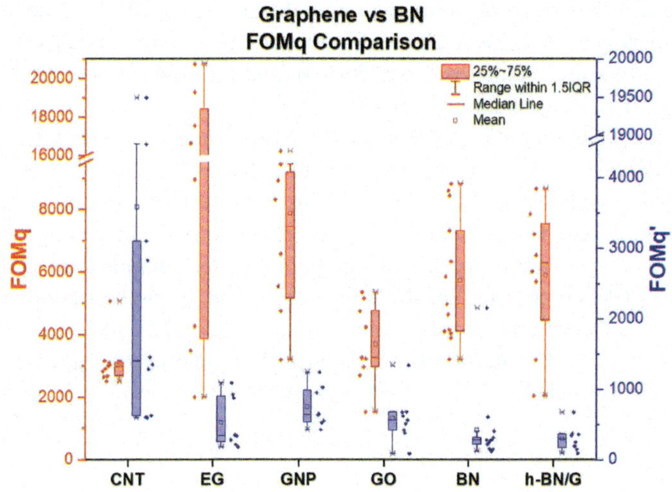

Fig. 18 Effect of filler in the overall $FOM_q/FOM_q^{'}$ metrics

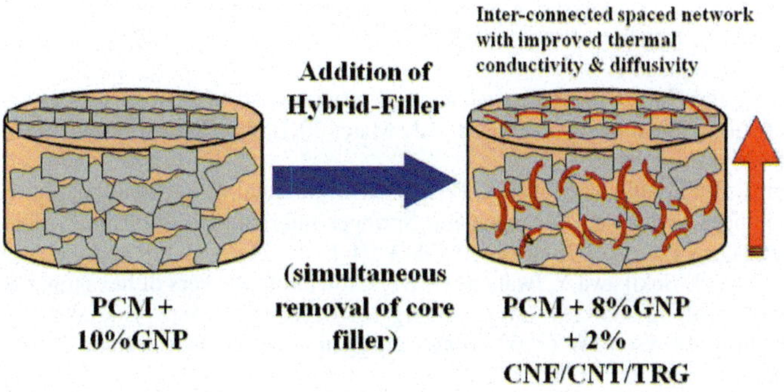

Fig. 19 Schematic of the proposed effect of the hybrid nanofillers on composite properties [116]

6 Concluding Remarks

Despite its relatively early stage of research development, BN-based PCM nanocomposites have immense potential to revolutionize the future of renewable energy systems. However, the bottleneck remains the optimization of the PCM properties to suit the desired end application. The focus of research should remain on improving the thermal conductivity of these PCMs as it remains the main bottleneck in their practical application as TES or TIM. While carbon-based nanofillers have been preferred for enhancing thermal properties, lately, BN has emerged as a formidable challenger to the status quo. It offers relatively high thermal conductivity, enough to bring it at par with composites based on carbon-based nanofillers.

While different strategies have been discussed in earlier sections, a focus on structured 3D fillers or hybrid combinations of different nanofillers should be taken up by researchers. The synergistic effect of hybrid combinations could prove extremely useful in addressing the thermal conductivity and have added benefits like flame-retardancy for BN-based composites.

Moreover, we observed a lack of standardization when reporting material properties, especially for PCM nanocomposites. A few studies may have provided detailed characterizations; however, most focus on just the thermal conductivity and basic properties like latent heat or melting temperature, to name a couple. Very few provide the entire set of material properties that could aid in determining the FOMq parameter for the PCM composite. We strongly suggest that researchers adopt a standard material characterization protocol to enable others to assess the material's viability in practical settings or via a figure-of-merit approach. A detailed outline of the protocol is available in Ref. [11].

References

1. Akovali G (2001) Handbook of composite fabrication, Rapra Technology, Shrewsbury
2. Matthews FL, Rawlings RD (1999) Composite materials: engineering and science. Elsevier Science, Woodhead P
3. Geng D, Yang HY (2018) Recent advances in growth of novel 2D materials: beyond graphene and transition metal dichalcogenides. Adv Mater 30:1800865. https://doi.org/10.1002/adma.201800865
4. Saleem J, Bin Shahid U, McKay G (2018) Environmental nanotechnology. In: Handbook of Environmental Materials Management. Springer International Publishing, Cham, pp 1–32. https://doi.org/10.1007/978-3-319-58538-3_94-1
5. Ishii T, Sato T, Sekikawa Y, Iwata M (1981) Growth of whiskers of hexagonal boron nitride. J Cryst Growth 52:285–289. https://doi.org/10.1016/0022-0248(81)90206-2
6. Niedenzu K, Dawson JW (1965) Boron-nitrogen compounds. Springer Berlin Heidelberg. https://doi.org/10.1007/978-3-642-85826-0
7. Balmain WH (1842) Bemerkungen über die Bildung von Verbindungen des Bors und Siliciums mit Stickstoff und gewissen Metallen. J Für Prakt Chemie 27:422–430. https://doi.org/10.1002/prac.18420270164
8. Wentorf RH (1957) Cubic form of boron nitride. J Chem Phys 26:956–956. https://doi.org/10.1063/1.1745964

9. Jansen M (ed) (2002) High performance non-oxide ceramics II. Springer Berlin Heidelberg. https://doi.org/10.1007/3-540-45623-6
10. Haubner R, Wilhelm M, Weissenbacher R, Lux B (2002) Boron nitrides—properties, synthesis and applications. Springer Berlin Heidelberg, pp 1–45. https://doi.org/10.1007/3-540-45623-6_1
11. Bin Shahid U, Abdala A (2021) A critical review of phase change material composite performance through figure-of-merit analysis: Graphene versus boron nitride. Energy Storage Mater 34:365–387. https://doi.org/10.1016/j.ensm.2020.10.004
12. Yin J, Li J, Hang Y, Yu J, Tai G, Li X, Zhang Z, Guo W (2016) Boron nitride nanostructures: fabrication, functionalization and applications. Small 12:2942–2968. https://doi.org/10.1002/smll.201600053
13. Wickramaratne D, Weston L, Van de Walle CG (2018) Monolayer to bulk properties of hexagonal boron nitride. J Phys Chem C 122:25524–25529. https://doi.org/10.1021/acs.jpcc.8b09087
14. Ambekar RS, Deshmukh A, Suárez-Villagrán MY, Das R, Pal V, Dey S, Miller JH, Machado LD, Kumbhakar P, Tiwary CS (2020) 2D hexagonal boron nitride-coated cotton fabric with self-extinguishing property. ACS Appl Mater Interfaces acsami.0c12647. https://doi.org/10.1021/acsami.0c12647
15. Seyferth D (1984) Gmelin handbook of inorganic chemistry, 8th edn. B. Boron Compounds, Organometallics 3:339–339. https://doi.org/10.1021/om00080a901
16. Lipp A, Schwetz KA, Hunold K (1989) Hexagonal boron nitride: fabrication, properties and applications. J Eur Ceram Soc 5:3–9. https://doi.org/10.1016/0955-2219(89)90003-4
17. Jiang P, Qian X, Yang R, Lindsay L (2018) Anisotropic thermal transport in bulk hexagonal boron nitride. Phys Rev Mater 2:064005. https://doi.org/10.1103/PhysRevMaterials.2.064005
18. Sarvar F, Whalley D, Conway P (2006) Thermal interface materials—a review of the state of the art. In: 2006 1st electronic systemintegration technology conference. IEEE, pp 1292–1302.https://doi.org/10.1109/ESTC.2006.280178
19. Sharma SD, Kitano H, Sagara K (2004) Phase change materials for low temperature solar thermal applications. Res Rep Fac Eng Mie Univ 29:31–64. https://www.researchgate.net/publication/241091651_Phase_Change_Materials_for_Low_Temperature_Solar_Thermal_Applications.
20. Lane GA (1983) Solar heat storage: latent heat materials. United States
21. Abhat A (1983) Low temperature latent heat thermal energy storage: heat storage materials. Sol Energy 30:313–332. https://doi.org/10.1016/0038-092X(83)90186-X
22. Kaygusuz K (1999) The viability of thermal energy storage. Energy Sources 21:745–755. https://doi.org/10.1080/00908319950014489
23. Kaygusuz K (2000) Ahmet sari, energy and exergy calculations of latent heat energy storage systems. Energy Sources 22:117–126. https://doi.org/10.1080/00908310050014090
24. El-Dessouky H, Al-Juwayhel F (1997) Effectiveness of a thermal energy storage system using phase-change materials. Energy Convers Manag 38:601–617.https://doi.org/10.1016/S0196-8904(96)00072-6
25. Dinçer İ, Rosen MA (2010) Thermal energy storage. Wiley. https://doi.org/10.1002/9780470970751
26. Bin Shahid U, Siddharth K, Shao M (2021) Electrifying the nitrogen cycle: an electrochemical endeavor. Curr Opin Electrochem 30:100790. https://doi.org/10.1016/j.coelec.2021.100790
27. Bin Shahid U, Bicer Y, Ahzi S, Abdala A (2019)Thermodynamic assessment of an integrated renewable energy multigeneration system including ammonia as hydrogen carrier and phase change material energy storage. Energy Convers Manag 198:111809. https://doi.org/10.1016/j.enconman.2019.111809
28. Yao Y, Wang J, Bin Shahid U, Gu M, Wang H, Li H, Shao M (2020) Electrochemical synthesis of ammonia from nitrogen under mild conditions: current status and challenges. Electrochem Energy Rev https://doi.org/10.1007/s41918-019-00061-3
29. Hu H, Zhao Z, Wan W, Gogotsi Y, Qiu J (2013) Ultralight and highly compressible graphene aerogels. Adv Mater 25:2219–2223. https://doi.org/10.1002/adma.201204530

30. Tsakalakos L (2010) Introduction to photovoltaic physics, applications, and technologies. In: Nanotechnology Photovoltaics, CRC Press, pp 1–48. https://doi.org/10.1201/9781420076752-c1
31. Radziemska E, Klugmann E (2002) Thermally affected parameters of the current–voltage characteristics of silicon photocell. Energy Convers Manag 43:1889–1900. https://doi.org/10.1016/S0196-8904(01)00132-7
32. Skoplaki E, Palyvos JA (2009) On the temperature dependence of photovoltaic module electrical performance: a review of efficiency/power correlations. Sol Energy 83:614–624. https://doi.org/10.1016/j.solener.2008.10.008
33. Hanafi MZM, Ismail FS, Rosli R (2015) Radial plate fins heat sink model design and optimization. In: 2015 10th Asian control conference. IEEE, pp 1–5.https://doi.org/10.1109/ASCC.2015.7244448
34. Owais M, Zhao J, Imani A, Wang G, Zhang H, Zhang Z (2019) Synergetic effect of hybrid fillers of boron nitride, graphene nanoplatelets, and short carbon fibers for enhanced thermal conductivity and electrical resistivity of epoxy nanocomposites. Compos Part A Appl Sci Manuf 117:11–22. https://doi.org/10.1016/j.compositesa.2018.11.006
35. Owais M, Javed MH, Akram MZ, Paxton WF, Akhatov IS, Abaimov SG (2021) Review—recent advances in thermally conductive paper-like films. ECS J Solid State Sci Technol 10:033001. https://doi.org/10.1149/2162-8777/abea5b
36. Zhou Y, Wu S, Long Y, Zhu P, Wu F, Liu F, Murugadoss V, Winchester W, Nautiyal A, Wang Z, Guo Z (2020) Recent advances in thermal interface materials. ES Mater Manuf 7(7):4–24. https://doi.org/10.30919/esmm5f717
37. Wang B, Cunning BV, Kim NY, Kargar F, Park S, Li Z, Joshi SR, Peng L, Modepalli V, Chen X, Shen Y, Seong WK, Kwon Y, Jang J, Shi H, Gao C, Kim G, Shin TJ, Kim K, Kim J, Balandin AA, Lee Z, Ruoff RS (2019) Ultrastiff, strong, and highly thermally conductive crystalline graphitic films with mixed stacking order. Adv Mater 31:1903039. https://doi.org/10.1002/adma.201903039
38. Peng L, Xu Z, Liu Z, Guo Y, Li P, Gao C (2017) Ultrahigh thermal conductive yet superflexible graphene films. Adv Mater 29:1700589. https://doi.org/10.1002/adma.201700589
39. Feng C-P, Chen L-B, Tian G-L, Bai L, Bao R-Y, Liu Z-Y, Ke K, Yang M-B, Yang W (2020) Robust polymer-based paper-like thermal interface materials with a through-plane thermal conductivity over 9 Wm−1K−1. Chem Eng J 392:123784. https://doi.org/10.1016/j.cej.2019.123784
40. Wu D, Wu L, Zhou W, Sun Y, Zhang M (2010) Relations between the aspect ratio of carbon nanotubes and the formation of percolation networks in biodegradable polylactide/carbon nanotube composites. J Polym Sci Part B Polym Phys 48:479–489. https://doi.org/10.1002/polb.21909
41. Sarbu I, Sebarchievici C (2018) A comprehensive review of thermal energy storage. Sustainability 10:191. https://doi.org/10.3390/su10010191
42. Jiang F, Cui X, Song N, Shi L, Ding P (2020) Synergistic effect of functionalized graphene/boron nitride on the thermal conductivity of polystyrene composites. Compos Commun 20:100350. https://doi.org/10.1016/j.coco.2020.04.016
43. Han W, Ge C, Zhang R, Ma Z, Wang L, Zhang X (2019) Boron nitride foam as a polymer alternative in packaging phase change materials: synthesis, thermal properties and shape stability. Appl Energy 238:942–951. https://doi.org/10.1016/j.apenergy.2019.01.153
44. Qian Z, Shen H, Fang X, Fan L, Zhao N, Xu J (2018) Phase change materials of paraffin in h-BN porous scaffolds with enhanced thermal conductivity and form stability. Energy Build 158:1184–1188. https://doi.org/10.1016/j.enbuild.2017.11.033
45. Sun N, Sun J, Zeng X, Chen P, Qian J, Xia R, Sun R (2018) Hot-pressing induced orientation of boron nitride in polycarbonate composites with enhanced thermal conductivity. Compos Part A Appl Sci Manuf 110:45–52. https://doi.org/10.1016/j.compositesa.2018.04.010
46. Suri AK, Subramanian C, Sonber JK, Murthy TSRC (2010) Synthesis and consolidation of boron carbide: a review. Int Mater Rev 55:4–40. https://doi.org/10.1179/095066009X12506721665211

47. Wan Y-J, Li G, Yao Y-M, Zeng X-L, Zhu P-L, Sun R (2020) Recent advances in polymer-based electronic packaging materials. Compos Commun 19:154–167. https://doi.org/10.1016/j.coco.2020.03.011
48. Liu B, Li Y, Fei T, Han S, Xia C, Shan Z, Jiang J (2020) Highly thermally conductive polystyrene/polypropylene/boron nitride composites with 3D segregated structure prepared by solution-mixing and hot-pressing method. Chem Eng J 385:123829. https://doi.org/10.1016/j.cej.2019.123829
49. Liang D, Ren P, Ren F, Jin Y, Wang J, Feng C, Duan Q (2020) Synergetic enhancement of thermal conductivity by constructing BN and AlN hybrid network in epoxy matrix. J Polym Res 27:212. https://doi.org/10.1007/s10965-020-02193-3
50. Wang Z-G, Liu W, Liu Y-H, Ren Y, Li Y-P, Zhou L, Xu J-Z, Lei J, Li Z-M (2020) Highly thermal conductive, anisotropically heat-transferred, mechanically flexible composite film by assembly of boron nitride nanosheets for thermal management. Compos Part B Eng 180:107569. https://doi.org/10.1016/j.compositesb.2019.107569
51. Shen W, Wu W, Liu C, Wang Z, Huang Z (2020) Achieving a high thermal conductivity for segregated BN/PLA composites via hydrogen bonding regulation through cellulose network. Polym Adv Technol 31:1911–1920. https://doi.org/10.1002/pat.4916
52. Zeng X, Ye L, Yu S, Li H, Sun R, Xu J, Wong C-P (2015) Artificial nacre-like papers based on noncovalent functionalized boron nitride nanosheets with excellent mechanical and thermally conductive properties. Nanoscale 7:6774–6781. https://doi.org/10.1039/C5NR00228A
53. Hu Z, Wang S, Chen G, Zhang Q, Wu K, Shi J, Liang L, Lu M (2018) An aqueous-only, green route to exfoliate boron nitride for preparation of high thermal conductive boron nitride nanosheet/cellulose nanofiber flexible film. Compos Sci Technol 168:287–295. https://doi.org/10.1016/j.compscitech.2018.09.020
54. Zeng X, Yao Y, Gong Z, Wang F, Sun R, Xu J, Wong C-P (2015) Ice-templated assembly strategy to construct 3D boron nitride nanosheet networks in polymer composites for thermal conductivity improvement. Small 11:6205–6213. https://doi.org/10.1002/smll.201502173
55. Hu J, Huang Y, Yao Y, Pan G, Sun J, Zeng X, Sun R, Xu J-B, Song B, Wong C-P (2017) Polymer composite with improved thermal conductivity by constructing a hierarchically ordered three-dimensional interconnected network of BN. ACS Appl Mater Interfaces 9:13544–13553. https://doi.org/10.1021/acsami.7b02410
56. Chen X, Lim JSK, Yan W, Guo F, Liang YN, Chen H, Lambourne A, Hu X (2020) Salt template assisted bn scaffold fabrication toward highly thermally conductive epoxy composites. ACS Appl Mater Interfaces 12:16987–16996. https://doi.org/10.1021/acsami.0c04882
57. Gurijala A, Zando RB, Faust JL, Barber JR, Zhang L, Erb RM (2020) Castable and printable dielectric composites exhibiting high thermal conductivity via percolation-enabled phonon transport. Matter 2:1015–1024. https://doi.org/10.1016/j.matt.2020.02.001
58. Amin M, Putra N, Kosasih EA, Prawiro E, Luanto RA, Mahlia TMI (2017) Thermal properties of beeswax/graphene phase change material as energy storage for building applications. Appl Therm Eng 112:273–280. https://doi.org/10.1016/j.applthermaleng.2016.10.085
59. Chen X, Dong S, Kan Y, Zhou H, Hu J, Wang D (2016) 3D Cf/SiC-ZrC-ZrB2 composites fabricated via sol-gel process combined with reactive melt infiltration. J Eur Ceram Soc 36:3607–3613. https://doi.org/10.1016/j.jeurceramsoc.2016.04.011
60. Karthik M, Faik A, Blanco-Rodríguez P, Rodríguez-Aseguinolaza J, D'Aguanno B (2015) Preparation of erythritol–graphite foam phase change composite with enhanced thermal conductivity for thermal energy storage applications. Carbon NY 94:266–276.https://doi.org/10.1016/j.carbon.2015.06.075
61. Mehrali M, Latibari ST, Mehrali M, Indra Mahlia TM, Cornelis Metselaar HS, Naghavi MS, Sadeghinezhad E, Akhiani AR (2013) Preparation and characterization of palmitic acid/graphene nanoplatelets composite with remarkable thermal conductivity as a novel shape-stabilized phase change material. Appl Therm Eng 61:633–640. https://doi.org/10.1016/j.applthermaleng.2013.08.035
62. Qian T, Li J, Deng Y (2016) Pore structure modified diatomite-supported PEG composites for thermal energy storage. Sci Rep 6:32392. https://doi.org/10.1038/srep32392

63. Karaipekli A, Biçer A, Sarı A, Tyagi VV (2017) Thermal characteristics of expanded perlite/paraffin composite phase change material with enhanced thermal conductivity using carbon nanotubes. Energy Convers Manag 134:373–381. https://doi.org/10.1016/j.enconman.2016.12.053
64. Paoli C, Voyant C, Muselli M, Nivet M-L (2010) Forecasting of preprocessed daily solar radiation time series using neural networks. Sol Energy 84:2146–2160. https://doi.org/10.1016/j.solener.2010.08.011
65. Batch GL, Chen Y-T, Macoskot CW (1996) Capillary impregnation of aligned fibrous beds: experiments and model. J Reinf Plast Compos 15:1027–1051. https://doi.org/10.1177/073168449601501004
66. El Z, Bahy M, Zidan MF, Thabet MS (2008) J Hazard Mater 153(605):364–71 (n.d.)
67. Azadi P, Farnood R, Meier E (2010) Preparation of multiwalled carbon nanotube-supported nickel catalysts using incipient wetness method. J Phys Chem A 114:3962–3968. https://doi.org/10.1021/jp907403b
68. Gabelica Z, Charmot A, Vataj R, Soulimane R, Barrault J, Valange S (2009) Thermal degradation of iron chelate complexes adsorbed on mesoporous silica and alumina. J Therm Anal Calorim 95:445–454. https://doi.org/10.1007/s10973-008-9256-z
69. Jiang SP (2006) A review of wet impregnation—an alternative method for the fabrication of high performance and nano-structured electrodes of solid oxide fuel cells. Mater Sci Eng A 418:199–210. https://doi.org/10.1016/j.msea.2005.11.052
70. Yavari F, Fard HR, Pashayi K, Rafiee MA, Zamiri A, Yu Z, Ozisik R, Borca-Tasciuc T, Koratkar N (2011) Enhanced thermal conductivity in a nanostructured phase change composite due to low concentration graphene additives. J Phys Chem C 115:8753–8758. https://doi.org/10.1021/jp200838s
71. Luan Y, Yang M, Ma Q, Qi Y, Gao H, Wu Z, Wang G (2016) Introduction of an organic acid phase changing material into metal–organic frameworks and the study of its thermal properties. J Mater Chem A 4:7641–7649. https://doi.org/10.1039/C6TA01676F
72. Mitran R, Berger D, Munteanu C, Matei C (2015) Evaluation of different mesoporous silica supports for energy storage in shape-stabilized phase change materials with dual thermal responses. J Phys Chem C 119:15177–15184. https://doi.org/10.1021/acs.jpcc.5b02608
73. Fang H, Bai S-L, Wong CP (2017) Thermal, mechanical and dielectric properties of flexible BN foam and BN nanosheets reinforced polymer composites for electronic packaging application. Compos Part A Appl Sci Manuf 100:71–80. https://doi.org/10.1016/j.compositesa.2017.04.018
74. Ren H, Tang M, Guan B, Wang K, Yang J, Wang F, Wang M, Shan J, Chen Z, Wei D, Peng H, Liu Z (2017) Hierarchical graphene foam for efficient omnidirectional solar-thermal energy conversion. Adv Mater 29:1702590. https://doi.org/10.1002/adma.201702590
75. Ito Y, Tanabe Y, Han J, Fujita T, Tanigaki K, Chen M (2015) Multifunctional porous graphene for high-efficiency steam generation by heat localization. Adv Mater 27:4302–4307. https://doi.org/10.1002/adma.201501832
76. Xu X, Zhang Q, Hao M, Hu Y, Lin Z, Peng L, Wang T, Ren X, Wang C, Zhao Z, Wan C, Fei H, Wang L, Zhu J, Sun H, Chen W, Du T, Deng B, Cheng GJ, Shakir I, Dames C, Fisher TS, Zhang X, Li H, Huang Y, Duan X (2019) Double-negative-index ceramic aerogels for thermal superinsulation. Science 363:723–727. https://doi.org/10.1126/science.aav7304
77. Song L, Ci L, Lu H, Sorokin PB, Jin C, Ni J, Kvashnin AG, Kvashnin DG, Lou J, Yakobson BI, Ajayan PM (2010) Large scale growth and characterization of atomic hexagonal boron nitride layers. Nano Lett 10:3209–3215. https://doi.org/10.1021/nl1022139
78. Shi Y, Hamsen C, Jia X, Kim KK, Reina A, Hofmann M, Hsu AL, Zhang K, Li H, Juang Z-Y, Dresselhaus MS, Li L-J, Kong J (2010) Synthesis of few-layer hexagonal boron nitride thin film by chemical vapor deposition. Nano Lett 10:4134–4139. https://doi.org/10.1021/nl1023707
79. Ismach A, Chou H, Ferrer DA, Wu Y, McDonnell S, Floresca HC, Covacevich A, Pope C, Piner R, Kim MJ, Wallace RM, Colombo L, Ruoff RS (2012) Toward the controlled synthesis of hexagonal boron nitride films. ACS Nano 6:6378–6385. https://doi.org/10.1021/nn301940k

80. Chen J, Huang X, Zhu Y, Jiang P (2017) Cellulose nanofiber supported 3d interconnected bn nanosheets for epoxy nanocomposites with ultrahigh thermal management capability. Adv Funct Mater 27:1604754. https://doi.org/10.1002/adfm.201604754
81. Jiang F, Cui S, Song N, Shi L, Ding P (2018) Hydrogen Bond-regulated boron nitride network structures for improved thermal conductive property of polyamide-imide composites. ACS Appl Mater Interfaces 10:16812–16821. https://doi.org/10.1021/acsami.8b03522
82. Wu Y, Wang Z, Shen X, Liu X, Han NM, Zheng Q, Mai Y-W, Kim J-K (2018) Graphene/boron nitride-polyurethane microlaminates for exceptional dielectric properties and high energy densities. ACS Appl Mater Interfaces 10:26641–26652. https://doi.org/10.1021/acsami.8b08031
83. Han J, Du G, Gao W, Bai H (2019) An anisotropically high thermal conductive boron nitride/epoxy composite based on nacre-mimetic 3D network. Adv Funct Mater 29:1900412. https://doi.org/10.1002/adfm.201900412
84. Wu Y, Xue Y, Qin S, Liu D, Wang X, Hu X, Li J, Wang X, Bando Y, Golberg D, Chen Y, Gogotsi Y, Lei W (2017) BN nanosheet/polymer films with highly anisotropic thermal conductivity for thermal management applications. ACS Appl Mater Interfaces 9:43163–43170. https://doi.org/10.1021/acsami.7b15264
85. Yao Y, Zeng X, Wang F, Sun R, Xu J, Wong C-P (2016) Significant enhancement of thermal conductivity in bioinspired freestanding boron nitride papers filled with graphene oxide. Chem Mater 28:1049–1057. https://doi.org/10.1021/acs.chemmater.5b04187
86. Wu K, Fang J, Ma J, Huang R, Chai S, Chen F, Fu Q (2017) Achieving a collapsible, strong, and highly thermally conductive film based on oriented functionalized boron nitride nanosheets and cellulose nanofiber. ACS Appl Mater Interfaces 9:30035–30045. https://doi.org/10.1021/acsami.7b08214
87. Guo F, Shen X, Zhou J, Liu D, Zheng Q, Yang J, Jia B, Lau AKT, Kim J (2020) Highly thermally conductive dielectric nanocomposites with synergistic alignments of graphene and boron nitride nanosheets. Adv Funct Mater 30:1910826. https://doi.org/10.1002/adfm.201910826
88. Xue Y, Zhou X, Zhan T, Jiang B, Guo Q, Fu X, Shimamura K, Xu Y, Mori T, Dai P, Bando Y, Tang C, Golberg D (2018) Densely interconnected porous BN frameworks for multifunctional and isotropically thermoconductive polymer composites. Adv Funct Mater 28:1801205. https://doi.org/10.1002/adfm.201801205
89. Yao Y, Ye Z, Huang F, Zeng X, Zhang T, Shang T, Han M, Zhang W, Ren L, Sun R, Xu J-B, Wong C-P (2020) Achieving significant thermal conductivity enhancement via an ice-templated and sintered BN-SiC skeleton. ACS Appl Mater Interfaces 12:2892–2902. https://doi.org/10.1021/acsami.9b19280
90. Jo I, Pettes MT, Kim J, Watanabe K, Taniguchi T, Yao Z, Shi L (2013) Thermal conductivity and phonon transport in suspended few-layer hexagonal boron nitride. Nano Lett 13:550–554. https://doi.org/10.1021/nl304060g
91. Loeblein M, Tsang SH, Pawlik M, Phua EJR, Yong H, Zhang XW, Gan CL, Teo EHT (2017) High-density 3D-boron nitride and 3D-graphene for high-performance nano-thermal interface material. ACS Nano 11:2033–2044. https://doi.org/10.1021/acsnano.6b08218
92. Shamberger PJ (2016) Cooling capacity figure of merit for phase change materials. J Heat Transfer 138. https://doi.org/10.1115/1.4031252
93. Stefan J (1891) Ueber die Theorie der Eisbildung, insbesondere über die Eisbildung im Polarmeere. Ann Phys 278:269–286. https://doi.org/10.1002/andp.18912780206
94. GH B (1901) Die Partiellen Differentialgleichungen der mathematischen Physik Nach Riemann's Vorlesungen. Nature 63:390–392. https://doi.org/10.1038/063390a0
95. Carslaw HS, Jaeger JC (1959) Conduction of heat in solids. Clarendon Press, Oxford, UK
96. Hu H, Argyropoulos SA (1996) Mathematical modelling of solidification and melting: a review. Model Simul Mater Sci Eng 4:371–396. https://doi.org/10.1088/0965-0393/4/4/004
97. Gowda A, Tonapi S, Reitz B, Gensler G (2005) Choosing the right thermal interface material. Adv Packag 14–18

98. Fang X, Fan LW, Ding Q, Yao XL, Wu YY, Hou JF, Wang X, Yu ZT, Cheng GH, Hu YC (2014) Thermal energy storage performance of paraffin-based composite phase change materials filled with hexagonal boron nitride nanosheets. Energy Convers Manag 80:103–109. https://doi.org/10.1016/j.enconman.2014.01.016
99. Su D, Jia Y, Alva G, Tang F, Fang G (2016) Preparation and thermal properties of n–octadecane/stearic acid eutectic mixtures with hexagonal boron nitride as phase change materials for thermal energy storage. Energy Build 131:35–41. https://doi.org/10.1016/j.enbuild.2016.09.022
100. Yang J, Tang LS, Bao RY, Bai L, Liu ZY, Yang W, Xie BH, Yang MB (2016) An ice-templated assembly strategy to construct graphene oxide/boron nitride hybrid porous scaffolds in phase change materials with enhanced thermal conductivity and shape stability for light-thermal-electric energy conversion. J Mater Chem A 4:18841–18851. https://doi.org/10.1039/c6ta08454k
101. Yang J, Tang LS, Bao RY, Bai L, Liu ZY, Yang W, Xie BH, Yang MB (2017) Largely enhanced thermal conductivity of poly (ethylene glycol)/boron nitride composite phase change materials for solar-thermal-electric energy conversion and storage with very low content of graphene nanoplatelets. Chem Eng J 315:481–490. https://doi.org/10.1016/j.cej.2017.01.045
102. Zhang J, Wang X, Yu C, Li Q, Li Z, Li C, Lu H, Zhang Q, Zhao J, Hu M, Yao Y (2017) A facile method to prepare flexible boron nitride/poly(vinyl alcohol) composites with enhanced thermal conductivity. Compos Sci Technol 149:41–47. https://doi.org/10.1016/j.compscitech.2017.06.008
103. Yang J, Tang LS, Bao RY, Bai L, Liu ZY, Xie BH, Yang MB, Yang W (2018) Hybrid network structure of boron nitride and graphene oxide in shape-stabilized composite phase change materials with enhanced thermal conductivity and light-to-electric energy conversion capability. Sol Energy Mater Sol Cells 174:56–64. https://doi.org/10.1016/j.solmat.2017.08.025
104. Yu C, Gong W, Tian W, Zhang Q, Xu Y, Lin Z, Hu M, Fan X, Yao Y (2018) Hot-pressing induced alignment of boron nitride in polyurethane for composite films with thermal conductivity over 50 $Wm^{-1} K^{-1}$. Compos Sci Technol 160:199–207. https://doi.org/10.1016/j.compscitech.2018.03.028
105. Cui X, Ding P, Zhuang N, Shi L, Song N, Tang S (2015) Thermal conductive and mechanical properties of polymeric composites based on solution-exfoliated boron nitride and graphene nanosheets: a morphology-promoted synergistic effect. ACS Appl Mater Interfaces 7:19068–19075. https://doi.org/10.1021/acsami.5b04444
106. Jeong S-G, Lee J-H, Seo J, Kim S (2014) Thermal performance evaluation of bio-based shape stabilized PCM with boron nitride for energy saving. Int J Heat Mass Transf 71:245–250. https://doi.org/10.1016/j.ijheatmasstransfer.2013.12.017
107. Yang Z, Zhou L, Luo W, Wan J, Dai J, Han X, Fu K, Henderson D, Yang B, Hu L (2016) Thermally conductive, dielectric PCM-boron nitride nanosheet composites for efficient electronic system thermal management. Nanoscale 8:19326–19333. https://doi.org/10.1039/c6nr07357c
108. Shao L, Shi L, Li X, Song N, Ding P (2016) Synergistic effect of BN and graphene nanosheets in 3D framework on the enhancement of thermal conductive properties of polymeric composites. Compos Sci Technol 135:83–91. https://doi.org/10.1016/j.compscitech.2016.09.013
109. Lei C, Wu K, Wu L, Liu W, Du R, Chen F, Fu Q (2019) Phase change material with anisotropically high thermal conductivity and excellent shape stability due to its robust cellulose/BNNSs skeleton. J Mater Chem A 7:19364–19373. https://doi.org/10.1039/c9ta05067a
110. Wan L, Liu C, Cao D, Sun X, Zhu H (2020) High phase change enthalpy enabled by nanocellulose enhanced shape stable boron nitride aerogel. ACS Appl Polym Mater 1–9. https://doi.org/10.1021/acsapm.0c00445
111. Xie B, Li C, Chen J, Wang N (2020) Exfoliated 2D hexagonal boron nitride nanosheet stabilized stearic acid as composite phase change materials for thermal energy storage. Sol Energy 204:624–634. https://doi.org/10.1016/j.solener.2020.05.004

112. Xia Y, Cui W, Ji R, Huang C, Huang Y, Zhang H, Xu F, Huang P, Li B, Sun L (2020) Design and synthesis of novel microencapsulated phase change materials with enhancement of thermal conductivity and thermal stability: self-assembled boron nitride into shell materials. Colloids Surf A Physicochem Eng Asp 586:124225. https://doi.org/10.1016/j.colsurfa.2019.124225
113. Wang Y, Tang B, Zhang S (2013) Single-walled carbon nanotube/phase change material composites: sunlight-driven, reversible, form-stable phase transitions for solar thermal energy storage. Adv Funct Mater 23:4354–4360. https://doi.org/10.1002/adfm.201203728
114. Fan L-W, Fang X, Wang X, Zeng Y, Xiao Y-Q, Yu Z-T, Xu X, Hu Y-C, Cen K-F (2013) Effects of various carbon nanofillers on the thermal conductivity and energy storage properties of paraffin-based nanocomposite phase change materials. Appl Energy 110:163–172. https://doi.org/10.1016/j.apenergy.2013.04.043
115. Wang J, Xie H, Xin Z, Li Y, Chen L (2010) Enhancing thermal conductivity of palmitic acid based phase change materials with carbon nanotubes as fillers. Sol Energy 84:339–344. https://doi.org/10.1016/j.solener.2009.12.004
116. Shahid U, Ahzi S, Abdala A (2018) Effect of carbon-based hybrid nano-fillers on the thermal conductivity of PCM composites. IOP Conf. Ser. Mater. Sci. Eng. 380:012008. https://doi.org/10.1088/1757-899X/380/1/012008
117. Xu B, Li Z (2014) Paraffin/diatomite/multi-wall carbon nanotubes composite phase change material tailor-made for thermal energy storage cement-based composites. Energy 72:371–380. https://doi.org/10.1016/j.energy.2014.05.049
118. Kumaresan V, Velraj R, Das SK (2012) The effect of carbon nanotubes in enhancing the thermal transport properties of PCM during solidification. Heat Mass Transf 48:1345–1355. https://doi.org/10.1007/s00231-012-0980-3
119. Wang J, Xie H, Xin Z (2009) Thermal properties of paraffin based composites containing multi-walled carbon nanotubes. Thermochim Acta 488:39–42. https://doi.org/10.1016/j.tca.2009.01.022
120. Dao TD, Jeong HM (2015) Novel stearic acid/graphene core–shell composite microcapsule as a phase change material exhibiting high shape stability and performance. Sol Energy Mater Sol Cells 137:227–234. https://doi.org/10.1016/j.solmat.2015.02.009
121. Yuan K, Wang H, Liu J, Fang X, Zhang Z (2015) Novel slurry containing graphene oxide-grafted microencapsulated phase change material with enhanced thermo-physical properties and photo-thermal performance. Sol Energy Mater Sol Cells 143:29–37. https://doi.org/10.1016/j.solmat.2015.06.034
122. Qiao Z, Mao J, Enhanced thermal properties with graphene oxide in the urea-formaldehyde microcapsules containing paraffin PCMs. J Microencapsul 1–9. https://doi.org/10.1080/02652048.2016.1267811
123. Sari A, Karaipekli A (2009) Preparation, thermal properties and thermal reliability of palmitic acid/expanded graphite composite as form-stable PCM for thermal energy storage. Sol Energy Mater Sol Cells 93:571–576. https://doi.org/10.1016/j.solmat.2008.11.057
124. Dao TD, Jeong HM (2016) A Pickering emulsion route to a stearic acid/graphene core–shell composite phase change material. Carbon NY 99:49–57. https://doi.org/10.1016/j.carbon.2015.12.009
125. Li M, Wu Z, Tan J (2012) Properties of form-stable paraffin/silicon dioxide/expanded graphite phase change composites prepared by sol–gel method. Appl Energy 92:456–461. https://doi.org/10.1016/j.apenergy.2011.11.018
126. Zhong L, Zhang X, Luan Y, Wang G, Feng Y, Feng D (2014) Preparation and thermal properties of porous heterogeneous composite phase change materials based on molten salts/expanded graphite. Sol Energy 107:63–73. https://doi.org/10.1016/j.solener.2014.05.019
127. Mei D, Zhang B, Liu R, Zhang Y, Liu J (2011) Preparation of capric acid/halloysite nanotube composite as form-stable phase change material for thermal energy storage. Sol Energy Mater Sol Cells 95:2772–2777. https://doi.org/10.1016/j.solmat.2011.05.024
128. Yang J, Zhang E, Li X, Zhang Y, Qu J, Yu Z-Z (2016) Cellulose/graphene aerogel supported phase change composites with high thermal conductivity and good shape stability for thermal energy storage. Carbon NY 98:50–57. https://doi.org/10.1016/j.carbon.2015.10.082

129. Karaipekli A, Biçer A, Sarı A, Tyagi VV (2017) Thermal characteristics of expanded perlite/paraffin composite phase change material with enhanced thermal conductivity using carbon nanotubes. Energy Convers Manag 134:373–381. https://doi.org/10.1016/j.enconman.2016.12.053
130. Karaman S, Karaipekli A, Sar A, Biçer A, Sarı A, Biçer A (2011) Polyethylene glycol (PEG)/diatomite composite as a novel form-stable phase change material for thermal energy storage. Sol Energy Mater Sol Cells 95:1647–1653. https://doi.org/10.1016/j.solmat.2011.01.022
131. Duan Z, Zhang H, Sun L, Cao Z, Xu F, Zou Y, Chu H, Qiu S, Xiang C, Zhou H (2014) CaCl2·6H2O/Expanded graphite composite as form-stable phase change materials for thermal energy storage. J Therm Anal Calorim 115:111–117. https://doi.org/10.1007/s10973-013-3311-0
132. Mehrali MM, Latibari ST, Mehrali MM, Metselaar HSC, Silakhori M (2013) Shape-stabilized phase change materials with high thermal conductivity based on paraffin/graphene oxide composite. Energy Convers Manag 67:275–282. https://doi.org/10.1016/j.enconman.2012.11.023
133. Li M (2013) A nano-graphite/paraffin phase change material with high thermal conductivity. Appl Energy 106:25–30. https://doi.org/10.1016/j.apenergy.2013.01.031
134. Mehrali M, ten Elshof JE, Shahi M, Mahmoudi A (2021) Simultaneous solar-thermal energy harvesting and storage via shape stabilized salt hydrate phase change material. Chem Eng J 405:126624. https://doi.org/10.1016/j.cej.2020.126624
135. Zhu Y, Qin Y, Liang S, Chen K, Tian C, Wang J, Luo X, Zhang L (2019) Graphene/SiO2/n-octadecane nanoencapsulated phase change material with flower like morphology, high thermal conductivity, and suppressed supercooling. Appl Energy 250:98–108. https://doi.org/10.1016/j.apenergy.2019.05.021
136. Zhou Y, Li C, Wu H, Guo S (2020) Construction of hybrid graphene oxide/graphene nanoplates shell in paraffin microencapsulated phase change materials to improve thermal conductivity for thermal energy storage. Colloids Surf A Physicochem Eng Asp 597:124780. https://doi.org/10.1016/j.colsurfa.2020.124780
137. Zhao Q, Yang W, Li Y, He Z, Li Y, Zhou Y, Wang R, Fan J, Zhang K (2020) Multifunctional phase change microcapsules based on graphene oxide Pickering emulsion for photothermal energy conversion and superhydrophobicity. Int J Energy Res 44:4464–4474. https://doi.org/10.1002/er.5224
138. Yu J, Kong L, Wang H, Zhu H, Zhu Q, Su J (2019) A novel structure for heat transfer enhancement in phase change composite: rolled graphene film embedded in graphene foam. ACS Appl Energy Mater 2:1192–1198. https://doi.org/10.1021/acsaem.8b01752

Synthesis of Heteroatoms Including at Least One Boron Metal Complexes and Their Catalytic Applications

Md. Zafer Alam and Salman A. Khan

Abstract Over the past decades, the synthesis of boron metal complexes has been used in the field of medicinal chemistry, and material sciences as well as used as the catalyst for various organic reactions. The Present chapter provides the synthesis of heteroatoms including at least one boron metal complex including benzothiazole, hydroxy quinoline, Different types of Schiff bases, Pyrroyl, Benzodithiaborol, and Benzo-1,3,2-oxothiaborolane, etc. Further boron metal complexes have been reported as catalytic applications of various organic reactions such as the Conversion of sugar and agricultural waste into a useful product, reduction of ketone, hydrogenation, alkylation, Diels–Alder reaction and Strecker-type reaction.

Keywords Heteroatoms · Metal complxes · Catalytic applications

1 Introduction

Boron is the lightest metalloid that constitutes 0.001% of the weight of the earth's crust. Most of the physical and chemical properties of boron are lies between metallic and non-metallic characters. Its atomic number is 5, it is a group 13 and P-block element, and have 3 e^- in the valence shell and exits in a variable oxidation state is −I to +III. Boron rarely follows the octet rule but forms stable covalent compounds such as H_3BO_3 (hydrogen borate/Boric acid), B_4C (tetrabor/boron carbide), BN (azanylidyneborane/boron nitride), $B_4O_7Na_2.10H_2O$ (Sodium tetraborate decahydrate/borax), etc. Due to vacant p-orbital and incomplete octet e^-s in the valence shell some compounds of boron act as lewis acid, which forms dative bonds or coordinate bonds with a nucleophile. Therefore the hybridization and shape of the boron center changes from sp^2 (trigonal planar) to sp^3 (tetrahedral). Such as boron trihalide (BX_3) and its conjugate base $(BX_4)^-$. The boron word first came into existence from borax (tincal). Boron does not exist in elemental form in nature. Sir Humphry Davy [1] and

Md. Z. Alam · S. A. Khan (✉)
Physical Sciences Section (Chemistry), School of Sciences, Maulana Azad National Urdu University, Hyderabad 500032, Telangana, India
e-mail: sahmad_phd@yahoo.co.in

M. Jawaid and A. Khan (eds.), *Multifunctional Boron-Nitride Composites*, Composites Science and Technology, https://doi.org/10.1007/978-981-99-2866-8_2

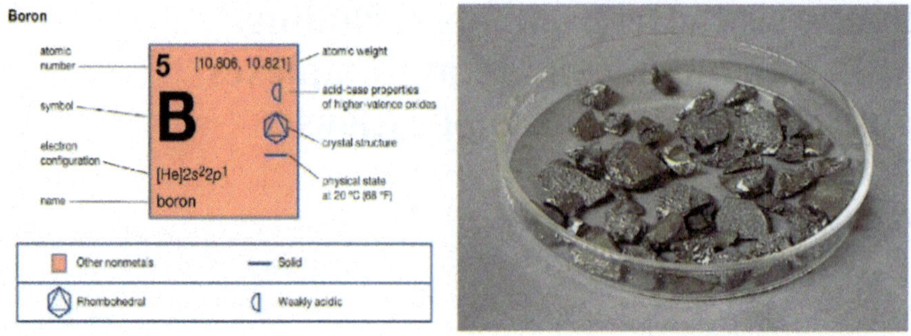

Fig. 1 Copyright, assignment point, and wikipedia

Gay-lussac and Thenard [2] (1808) first time isolate elemental boron from borate solution. Boron exists in two stable isotopes ^{11}B (80.1%) and ^{10}B (19.9%), it exists in different allotropes i.e. amorphous boron (brown powder), crystalline boron (silvery black) (polymorphs of crystalline boron are α-rhombohedral, β-rhombohedral, γ-orthorhombic and β-tetragonal). Crystalline boron is harder than amorphous boron, it is chemically inert and its boiling point is more than 2000 °C (Fig. 1).

The chemical feature of 'B' maximum resembles 'C' and 'Si' more than 'Al'. Boron doesn't react with O_2/air at room temperature but as the temperature increase, it starts to burn and produced B_2O_3 [3]. Some of the boron-containing compounds dimerize and also form isomers, Borane (BH_3) dimerize into diborane (B_2H_6), and some isomer is $B_{20}H_{26}$. There are many organoboron compounds are found, eg. $B(C_6H_5)_3$, etc. While some of the organoboron compounds help in the synthesis of an organic compound. Boron also forms stable compound as diboron tetrafluoride (B_2F_4), magnesium diboride (MgB_2), calcium hexaboride (CaB_6), carboranes ($C_2B_{10}H_{12}$) [3]. Many biomolecules have boron such as Aplasmomycine, Boromycine, Tartrolon, etc.

Boron is an essential micronutrient for plants and also important for animals. Boron helps in plant growth, cell wall formation and strengthening, transport of sugar to the growing region of plants, pollination, nitrogen fixation, etc. Boron deficiency in plants leads to stunting root growth, reduced pollen vitality, abnormal pollen grain, death of growing region meristematic tissue, effect pollination, etc.

Boron plays a major role in animal cell membrane function, enzymatic reaction, hormonal and mineral metabolism, etc. An imbalance of boron concentration in the human or animal body affects bone development, mineral (Ca, P, Mg, etc.) and energy metabolism, enzyme activation, cholesterol, etc. [4].

Boron is used in the formation of different things such as boron fiber (used in aerospace structures and goods for golf clubs and fishing rods, tungsten glass, etc.) [5, 6], and boronated fiberglass (electrical glass). Borosilicate glass (laboratory glassware and cookware) [7], boron carbide ceramic (neutron radiation absorbent) [8]. In metallurgy (boron steel) [9], detergent formulation and bleaching agent (as borax used in laundry and cleaning products, tooth bleaching) [10], insecticides (boric acid as an insecticide) [11], semiconductors [12], magnets [13], shielding and neutron absorber in nuclear reaction [14]. It has many other uses in the biological and pharmaceutical field (boric acid uses as an antiseptic, antifungal, and anti-viral) [15].

Similarly, BN (boron nitride) looks like an inorganic diamond since it is structurally analogous to diamond and graphite. BN nanostructures act as an insulator for electricity, resistant to oxidation/heat till 800 ^{0}C. BN nanostructures have many applications such as in the making of BN nanofibers [16], BN nanotubes [17–20], BN nanocapsules [21], BN nanowires [22], BN nanoparticles [23], BN nanohorns [24], BN nanosheets [25], films [26], etc.

There are many Schiff bases containing boron have been synthesized. Some of them are used as fluorescent chemosensors to recognize metal ions such as Rhodamine-BOIDIPY used to detect Fe (III) [27], some of them are used as anti-cancer and antioxidant [28], di-Schiff base use for investigation against cancer [29]. There is also some catalyzing agent used which is boron based on complex compound [30, 31], some Schiff bases used as organic light emitting devices [32, 33], some of the Schiff base act as e- transport materials [34], some of the Schiff base use as fluorescent bio labels [35] and some act as chemosensor [36].

2 Synthesis of Boron Metal Complex

Boron complex 'aza-BODIPY' (Scheme 1) has been synthesized by C.J. Ziegler and their co-worker (2016) from phthalocyanine precursor. The reaction happens between $PhB(OH)_2$ and 1,3-diiminoisoindoline (DII) (1:1) ratio in chlorobenzene, and reflux for 72 h. After the completion of the reaction via column chromatography pure compound was obtained, the silica was used as the solid phase and CH_2Cl_2 as eluent [37].

The boron complex compound (Scheme 2) has been designed by M.J. Ingleson et al. (2015) by the following reaction. When 4,5-dibromo-1,2,3-benzothiadiazole reacted with tributyl-(thiophen-2-yl)-stannane through Stille coupling reaction, the product (P) obtained, P on borylation with BBr_3 gives complex of boron [38].

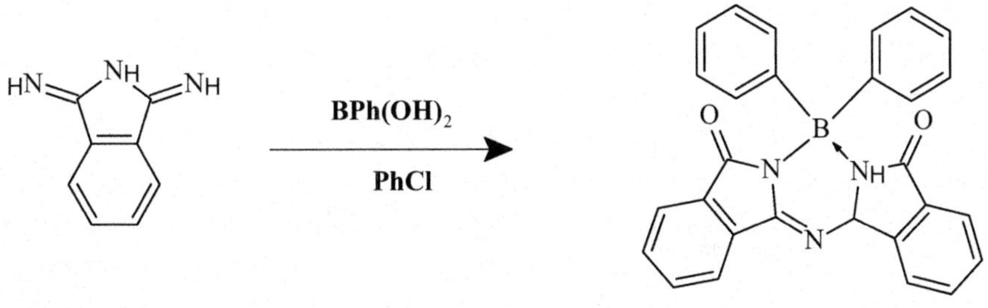

Scheme 1 Schematic diagram of aza-BODIPY

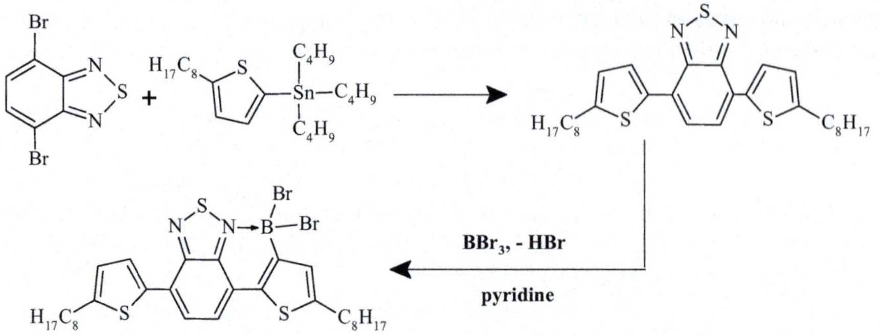

Scheme 2 Schematic diagram of boron complex by Stille coupling reaction

K.I. lee and their co-worker (2017) synthesized a boron (III) complex (Scheme 3) by the following reaction. When the reaction occurs between 8-hydroxyquinoline with Benzene-boronic acid in the presence of K_3PO_4 in 1,4-dioxane as a solvent, reflux for 20 h at 110 ^{0}C, 87% product yield [39].

P.J. Brothers (2011) synthesized a Boron containing pyrrolyl ligands (Scheme 4) by the oxidation of dipyrrane (dipyrromethane) in the presence of DDQ or chloranil. The obtained intermediate compound undergoes complexation with $BF_3{\cdot}OEt_2$ and formed BF_2 (dipyrrin) [40].

Boron containing fluorescent Schiff base (BOSCHIBAs) complex (Scheme 5) has been synthesized by B.M.M. Flores et al. (2017). Which was used in-vitro for cytoplasm staining dyes. The reaction occurred between 2-hydroxy-1-naphthaldehyde and 3-aminopyridine in acetonitrile and refluxed for 48 h. The yellow precipitate was obtained, washed with n-hexane, and a yellow solid Schiff base was obtained.

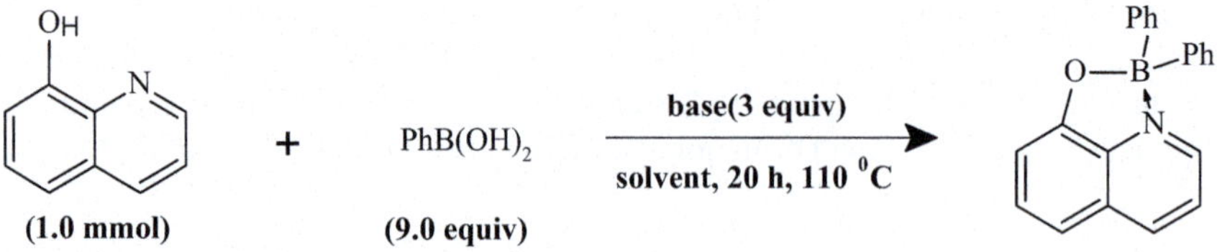

Scheme 3 Schematic structure of Boron (iii) complex

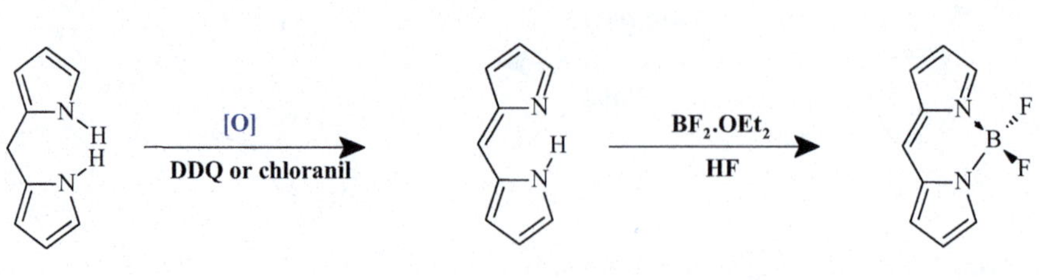

Scheme 4 Schematic structure of boron-containing pyrrolyl ligand

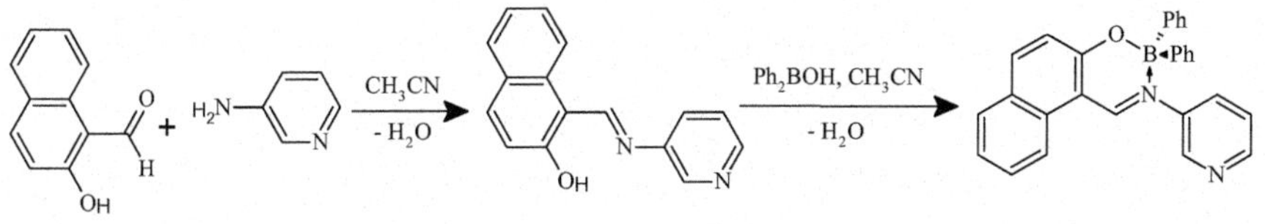

Scheme 5 Schematic structure of BOSCHIBAs

Obtained yellow Schiff base further reacts with diphenylboronic acid, which gives a yellow solid boron-containing Schiff base complex [41].

A. Kilic and their co-worker (2015) have synthesized a boron-containing complex compound (Scheme 6), When the mixture of 3,5-dibromosalicylaldehyde and 4-amino-2,2,6,6-tetramethylpiperidine in the presence of formic acid in 99% alcohol refluxed for 6 h, under N_2 atmosphere. The crude product after recrystallization gives a yellow solid Schiff base. The obtained Schiff base was further reacted with (diethyloxonio)-trifluoroborate ($BF_3.OEt_2$) in the presence of N,N-Diethylethanamine $(C_2H_5)_3N$, and refluxed for 4 h at 45 ^{0}C. The yellowish precipitates were obtained, which give a pure compound after recrystallization in CH_2Cl_2/ hexane [30].

A.N. Król and their co-worker (2017) have designed an acceptor-π-bridge-acceptor (A-π-A) boron-containing complex (Scheme 7) by the following reaction. When 2-Iodo-3-bromothiophene and tributyl-(1,3-thiazol-2-yl)stannane reacted through "Stille coupling reaction" in the presence of the catalytic amount of $Pd_2(dba)_3.CHCl_3$ and tri-2-furanylphosphine in toluene, and refluxed at 108 ^{0}C. 98% thienylthiazole (P_1) was formed. P_1 reacts with Mes_2BF according to the procedure proposed in the review paper [42], which gives di-mesityl-boryl (P_2). P_2 further reacts with nBuLi in DMF giving a derivative of aldehyde (P_3). Condensation reaction occurs between P_3 and CH-acids (malononitrile) in the presence of dry pyridine in dichloroethane, forming the final product i.e. (A-π-A) boron-containing complex [43].

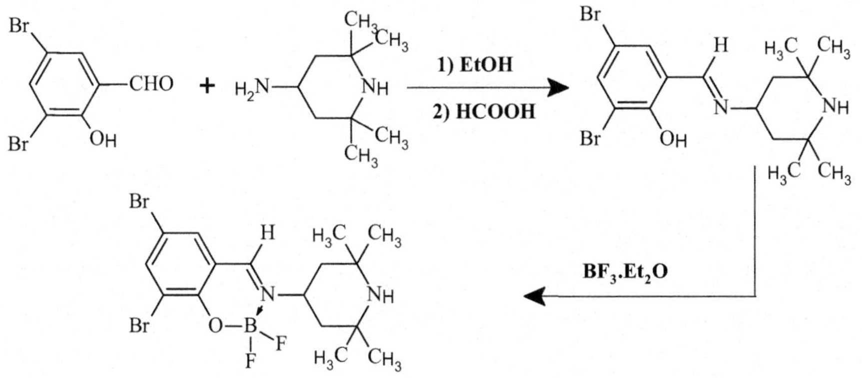

Scheme 6 Schematic structure of bromine-containing boron complex

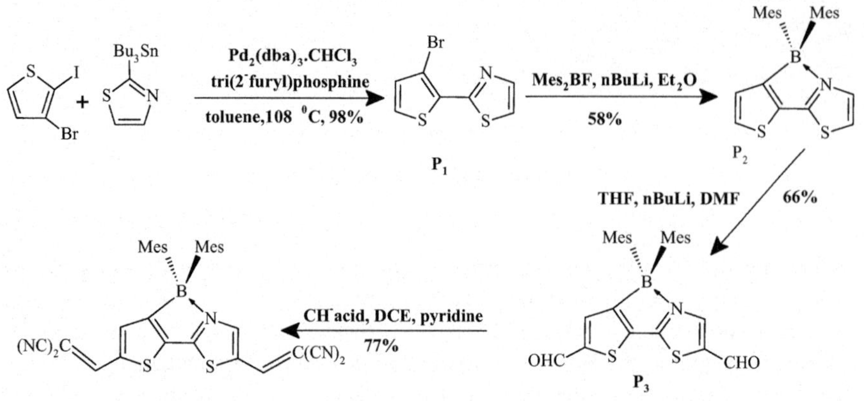

Scheme 7 Schematic diagram of (A-π-A) boron-containing complex

L. Liu and their co-worker (2015) synthesized a double NB bridging bipyridyl (BNBP) complex (Scheme 8) by the following reaction. Firstly, 2-chloro-3-nitropyridine undergoes dimerized in the presence of Cu powder in DMF at 150 ^{0}C, and the product (P1) was obtained. P1 undergoes a reduction in the presence of $SnCl_2.2H_2O$ and HCl and refluxed at 100 °C, the product (P2) was obtained. P2 upon alkylation with nBuLi in THF gives product (P3). P3 after reaction with $BF_3.Et_2O$/ Et_3N in dichloromethane at 50 °C gives 83% of a yellow solid BNBP-C4 complex. Pure complex compounds obtain through column chromatography [44].

K.N. Hadebe (2013) have been designed a boron-containing heteroatom compound (Benzo-1,3,2-thiazaborolanes) (Scheme 9a). The compound 1,2-aminothiophenol react with borane methyl disulfide complex in dry tetrahydrofuran

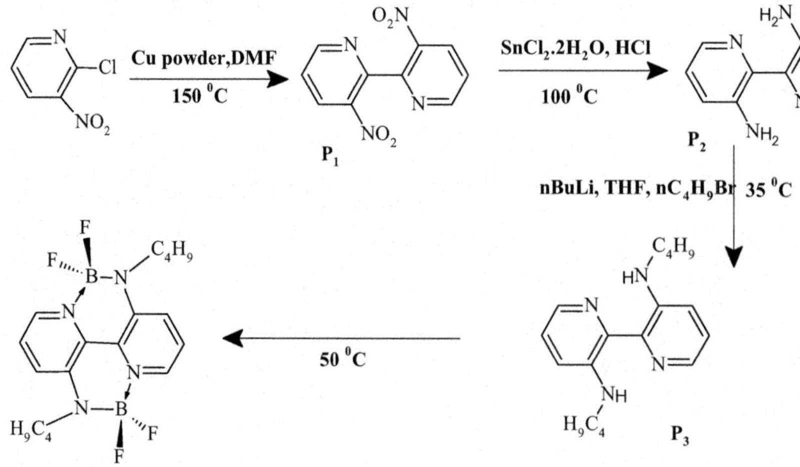

Scheme 8 Schematic indicates double NB bridging bipyridyl (BNBP) complex

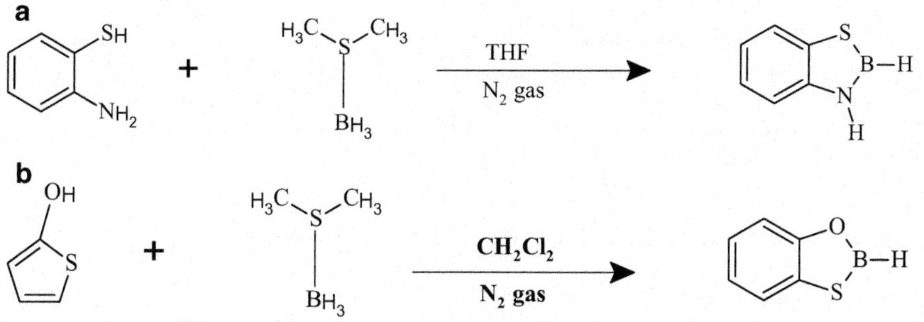

Scheme 9 **a** Schematic diagram of Benzo-1,3,2-thiazaborolanes. **b** Schematic diagram of Benzo-1,3,2-oxothiaborolane

(THF) under an N_2 atmosphere, reflux for 8 h at 40 °C. The milky yellowish solution obtained is Benzo-1,3,2-thiazaborolanes [45].

K.N. Hadebe (2013) have been designed a boron-containing heteroatom compound (Benzo-1,3,2-oxothiaborolane) (Scheme 9b). The compound 2-thiophenol reacts with borane-dimethyl sulfide complex in dry dichloromethane (CH_2Cl_2) under an N_2 atmosphere, reflux for 12 h at r t (25 °C). Benzo-1,3,2-oxothiaborolane was obtained [45].

A. Kilic and their co-worker (2011) have synthesized a sulfur-containing boron complex (Scheme 10). The reaction starts between compound 1 and $BF_3.OEt_2$ in anhydrous benzene and refluxed for 30 min in N_2 atmosphere. Then Et_3N was added to the solution and kept refluxed for 12 h, a Light green yellow precipitate was obtained, recrystallized with a suitable solvent, and a pure complex compound with a 70% yield was obtained [31]. Compound 1 was synthesized accordingly to the literature reported in [46].

A. Kilic et al. (2017) synthesized binuclear boron-fluoride complexes containing Hemi-salen/ salen ligand (Scheme 11). The reaction occurred between compound 1 and Boron trifluoride etherate ($BF_3.Et_2O$) in the presence of triethylamine (Et_3N) in

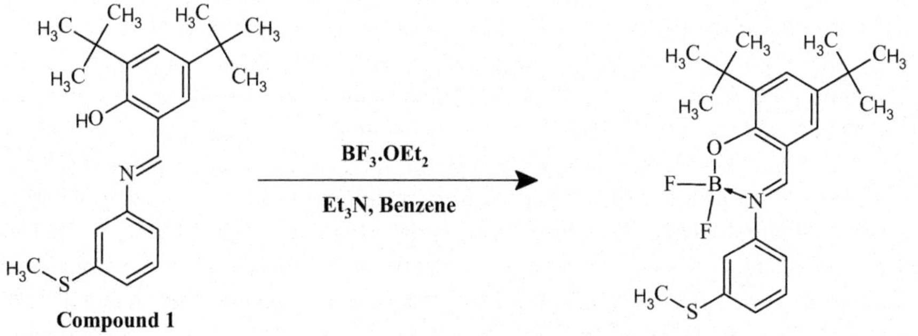

Scheme 10 Schematic diagram of sulfur-containing boron complex

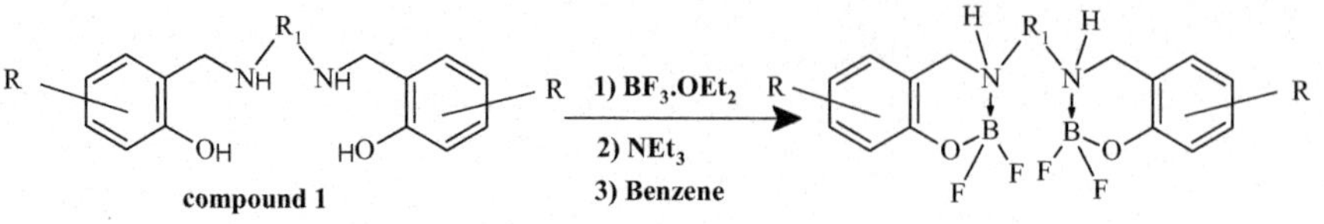

R=3,5-di-t-Bu , R$_1$=C$_8$H$_{18}$

Scheme 11 Schematic structure of binuclear boron-fluoride complexes

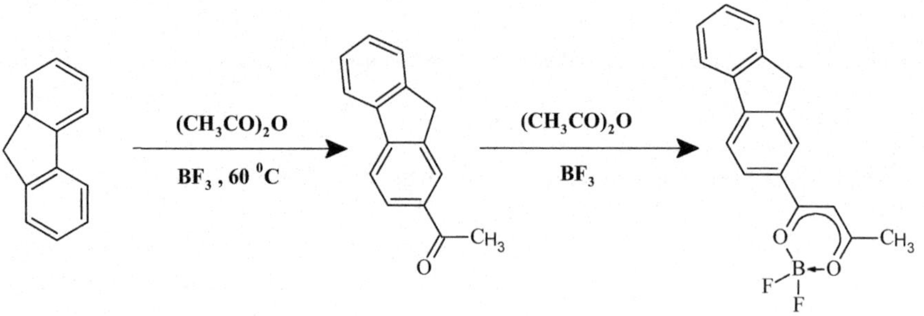

Scheme 12 The complex of boron

benzene. The yellow or white solid compounds were obtained with 86–79% yields [47].

B.V. Bukvetskii and their co-worker (2011) have synthesized a 2,2-difluoro-4-(9H-fluorene-2-yl)-6-methyl-1,3,2-dioxaborine (Scheme 12). When Fluorene reacts with gaseous BF_3 (boron trifluoride) in acetic anhydride, reflux is below 60 ^{0}C. The brown-colored precipitate was obtained, which that recrystallized with tetrahydrofuran and methylene chloride and gives a pure complex compound [48].

Fluorescent BODIPY-gluco-septanose and furanose (Scheme 13) have been synthesized by P.J. Brothers et al. (2016). The reaction happened between meso-p-tolyl substituted Fluorine-BODIPY with boron trichloride (BCl_3), giving Chlorine-BODIPY. The chlorine-BODIPY was added to D-glucose in anhydrous Ethanenitrile, and the solution color changed from dark red to orange. Through TLC analysis three new fluorescent green color products were shown under 365 nm UV light i.e. (a) 1:1 alpha-glucofuranose BODIPY, (b) 1:2 alpha-glucofuranose BODIPY and (c) 1:2 alpha-glucoseptanose BODIPY. The first two of them were nonpolar while the third one was polar. By column chromatography, all the products were separated [49].

K. Suzuki et al. (2009) reported "Tailor-made" multicolor fluorophores known as Keio Fluors (KFL) (Scheme 14), which was based on BODIPY (boron-dipyrromethene). When the derivative of 5-carboxylic-furopyrrole was reacted with the derivative of 5-formyl-furopyrrole in the presence of phosphoryl chloride in Perfluoroacetic acid and refluxed for 10–20 min at 50 °C. The reaction was completed, and obtained crude product was neutralized with $NaHCO_3$, and the precipitate was filtered out. $BF_3.Et_2O$ and triethylamine were added to the precipitate. The resultant

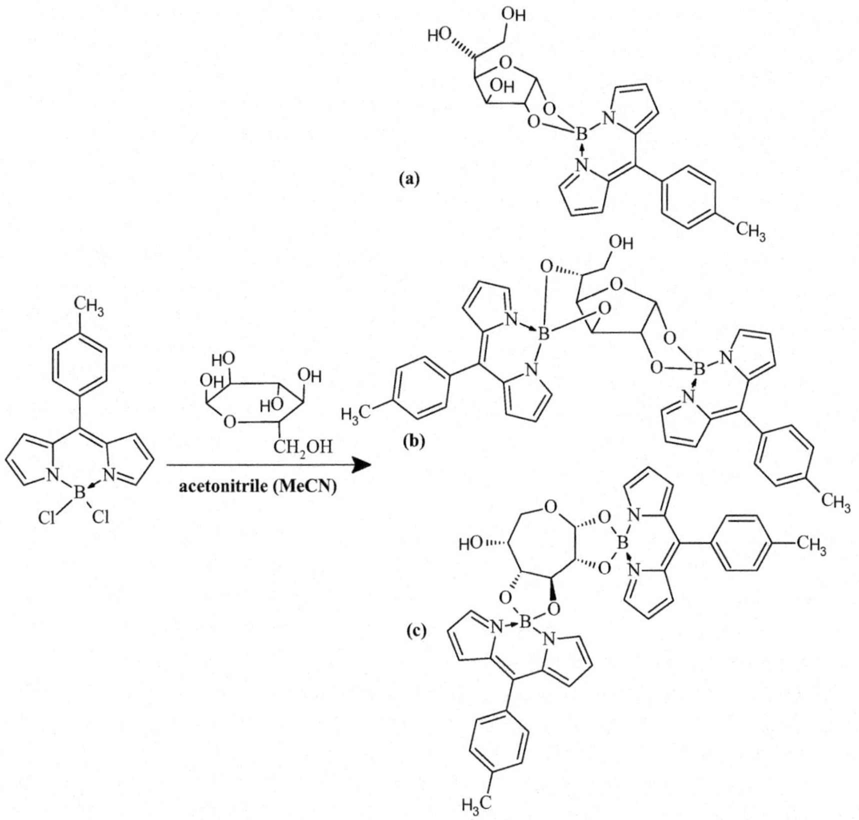

Scheme 13 Schematic structure of fluorescent BODIPY-gluco-septanose and furanose

mixture was dissolved in 1,1,2-trichloroethane and stirred for 15 min at 100 °C. The pure compound was obtained as a green metallic solid [50].

M.J. Hall and their co-worker (2014) synthesized an axially chiral BODIPYs complex (Scheme 15) by the following reaction. When 2,4-dimethyl pyrrole was reacted with 2-methyl benzoyl chloride in the presence of EtMgBr in Et_2O, the compound 1 i.e. (3,5-dimethylpyrrol-2-yl)-(o-tolyl)-methanone was obtained.

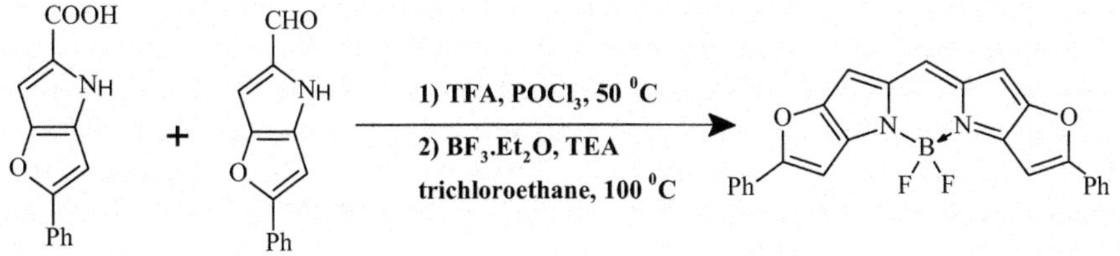

Scheme 14 The schematic structure of Keio flours (boron metal complex)

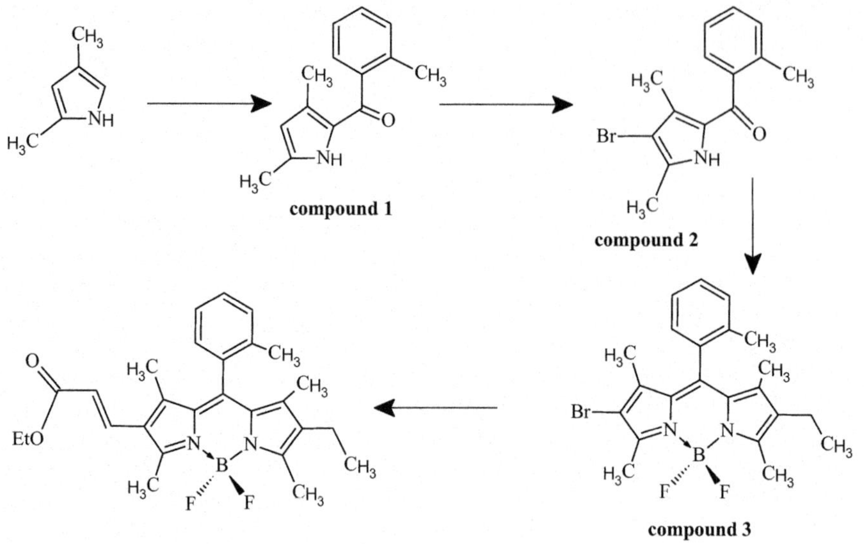

Scheme 15 Schematic diagram of axially chiral BODIPYs complex

Bromination occurs on the 4 positions of compound 1 in dichloromethane (CH_2Cl_2). Compound 2 i.e. (4-Bromo-3,5-dimethylpyrrol-2-yl)-(o-tolyl)-methanone was obtained. The condensation reaction takes place between compound 2 and i-Pr_2NEt, BF_3OEt_2, and compound 3 was obtained. Compound 3 further reacts with acrylic acid ethyl ester via Heck reaction in the presence of $Pd(OAc)_2$ (palladium(II) acetate), Et_3N (triethylamine), PPh_3 (triphenylphosphine) in DMF and refluxed it for 12 h at 100 °C, gives axially chiral BODIPYs compound [51].

Mono-O, B, N-strapped BODIPY complex (BDFO) (Scheme 16) have been synthesized by A. Harriman and their co-worker (2017) by the following reaction. Meso-tolyl BODIPY (BDF_2) react with free radical giving BDF_2OH (51%yield). Free radicals formed by the reaction of 2-phenyldiazonium hexafluorophosphate and ferrocene. The obtained BDF_2OH product was purified on silica gel. Pure BDF_2OH after intramolecular boron chelation in the presence of $SnCl_4$ gives BDFO the target product. Replacing fluorine with oxygen catalyze by either boron tribromide or aluminum trichloride [52].

R. Mazitschek and their co-worker (2014) have synthesized an alkoxy-fluoro-boron-dipyrromethenes core i.e. monoalkoxy-BODIPYs (Scheme 17), which were used in biomedical imaging techniques. The target compound was obtained in a two-step reaction. When derivatives of BODIPYs were reacted with trimethylsilyl trifluoromethanesulfonate (TMSOTf). Only one fluorine atom was substituted from BODIPYs and produced BODIPYs-OTF borenium intermediate. Borenium intermediate was reacted with alcohols giving monoalkoxy-BODIPYs complex without forming a dialkoxy-BODIPYs [53].

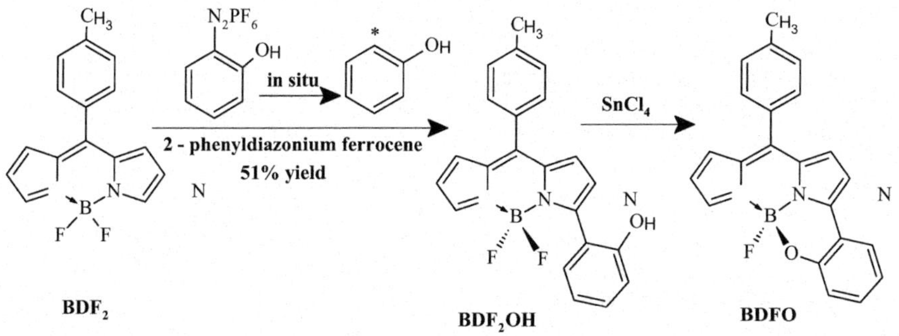

Scheme 16 Schematic represents mono-O, B, N-strapped BODIPY complex (BDFO)

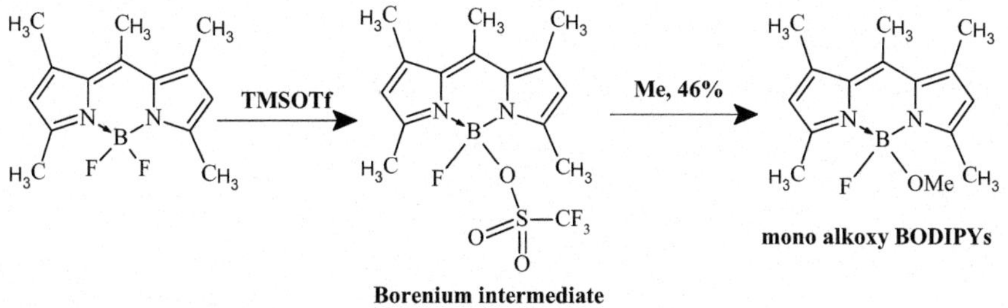

Scheme 17 Schematic indicates monoalkoxy-BODIPYs complex

G.W. Bauman and their co-worker (2012) have synthesized diarylborinic-8-oxyquinolates (Scheme 18) via bimetallic boron-lithium intermediates. Firstly p-Diiodo-benzene was reacted with n-Butyllithium in Et_2O at −78 °C giving 4-phenyllithium. Which was further reacted with diethyl-phenyl boronic acid ($PhB(OEt)_2$) giving iodinated diarylborinic "ate" complex i.e. ate-1. The Ate-1 undergoes halogen-lithium exchange with t-BuLi giving "ate-Li" as an intermediate. Ate-Li intermediate first hydrolyzed then reacted with 8-hydroxyquinoline and formed the target complex compound [54].

K.N. Jarzembska and their co-worker (2017) have synthesized a cheap and modifiable organic luminescent dye (Scheme 19) containing two boron atoms. When o-phenylenediboronic acid was reacted with 8-hydroxyquinoline gives spirocyclic benzooxadiborole. The obtained compound was recrystallized with diethyl ether and the pure compound was obtained [55].

Benzodithiaborole (Scheme 20) was designed and synthesized by D.A. Dixon et al. (2009). The reaction occurred between benzenedithiol and polyborazylene (PB) in THF and refluxed the mixture for 12 h. "Polyborazylene is the byproduct of ammonia-borane". Benzodithiaborole was formed with the release of H_2 gas. In benzodithiaborole B-S bond was formed which was weaker than the analogous B-O bond. Therefore benzodithiaborole in the presence of a catalyst (Bu_3SnH) reduced into ammonia borane (AB) [56].

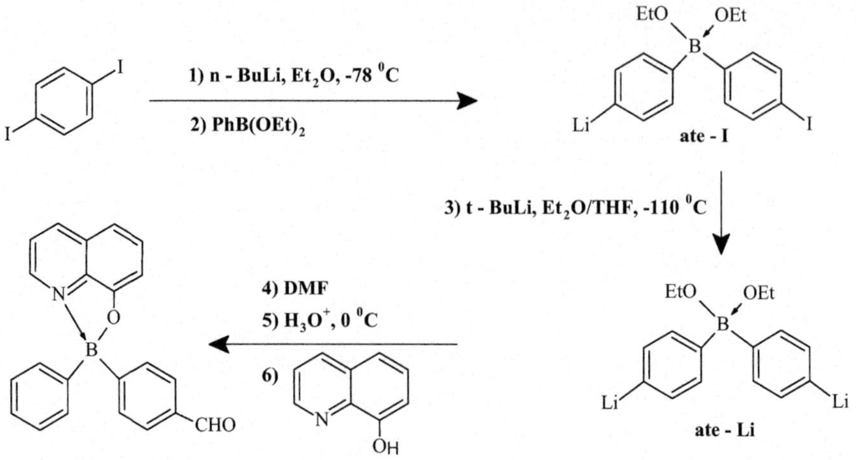

Scheme 18 Schematic diagram of diarylborinic 8-oxyquinolates

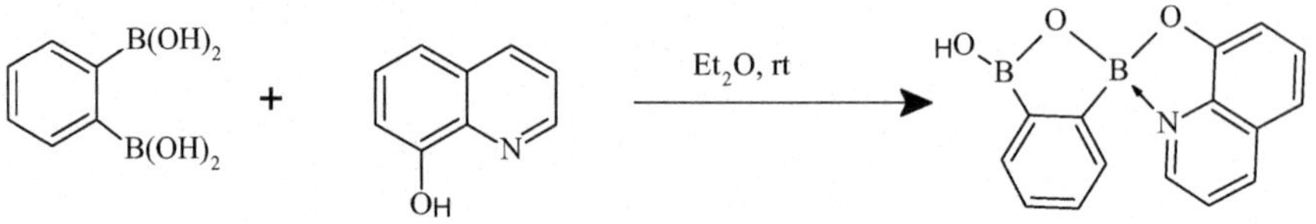

Scheme 19 Schematic represents two boron-containing luminescent dye

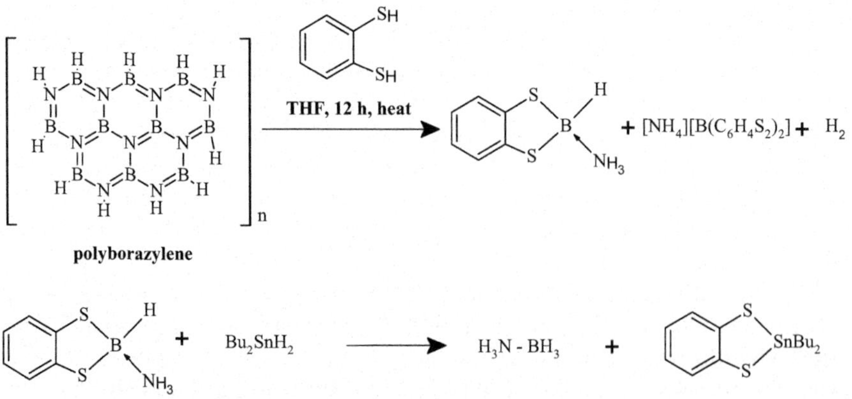

Scheme 20 Schematic indicates benzodithiaborole

3 Catalytic Applications of the Boron Metal Complex Are as

Since catalyst increases the kinetics of the reaction without disturbing the thermodynamics of the substrates and products, that's why catalysts are used in material science, chemical science, and biological science. Catalysts are found in three different forms, such as homogeneous catalysts, heterogeneous catalysts, and biological catalysts.

(a) Homogeneous catalysts are generally organometallic complexes with transition metals at the center. It occurs in the same phase as of reactant. In this catalyst transition-metal acts as an active site.
(b) Heterogeneous catalysts are a common type of catalyst used in the chemical industry, this type of catalyst occurs in different phases as of reactant.
(c) Biocatalysts which is enzymatic catalysts that enhance the chemical reaction in living organism.

Many boron-based catalysts have been synthesized which give more attention because it is cheaper than other noble metals catalyst. Boron-based catalysts are stable and work under both acidic medium and basic medium. Boron-based catalysts are synthesized by different techniques such as simple wet synthesis [57], hydrothermal process [58], deposition–precipitation method [59], sol–gel condensation method [60], a templating method [61], incipient wetness impregnation method [62], etc. $HB(C_6F_5)_2$ is a boron-based homogeneous catalyst that is used for Z-selective hydrogenation of hexyne to hexane with 87% yield [63]. Similarly metal hydride complex is a heterogeneous catalyst that can be used for the reduction of CO_2 [64].

4 Conversion of Sugar and Agricultural Waste into a Useful Product

M.G. de Luna and their co-worker (2020) used boron-doped biochar (BA/BCB) as a catalyst to convert the sugar (xylose, glucose), agricultural waste (wheat and rice straw, corncob) for the formation of furfural and 5-hydroxymethylfurfural (HMF) (Fig. 2). They used many other catalysts but BA/BCB gives the maximum yield of 4.14%–10.41%, respectively [65].

5 Reduction of Aromatic Ketones

Chelate boron hybrid complex (Fig. 3) a catalytic agent was used by A. Kilic and their co-workers (2015) for the reduction of aromatic ketones into corresponding alcohols. The ($L_{(1-3)}BF_3$) fluorine boron complexes and ($L_{(1-3)}BPh_3$) phenyl boron

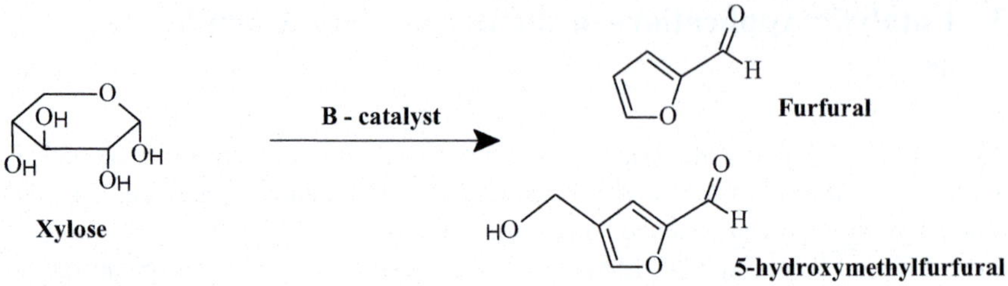

Fig. 2 BA/BCB as a catalyst; Copyright, 2020, Bioresource Technology Reports, Elsevier

complexes [30] are the catalyst, were used to examine the transfer hydrogenation of aromatic ketones by isopropyl alcohol (iso-PrOH). When catalyst and acetophenone (1:100) molar ratio were added with iso-PrOH in NaOH solution and reflux at 82 °C, the hydrogen of an iso-PrOH moved to acetophenone, and 1-phenylethyl alcohol was obtained [30].

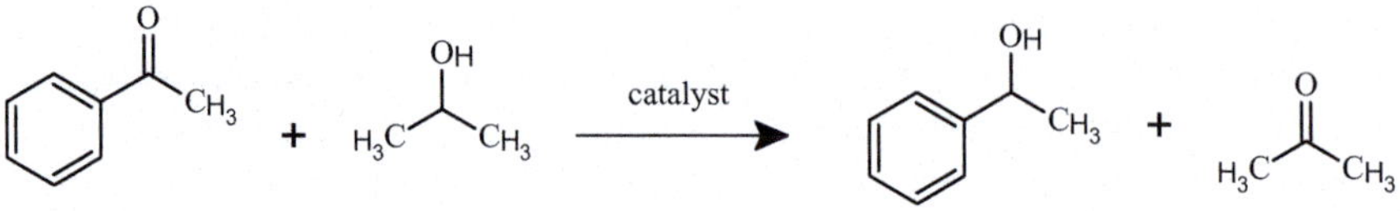

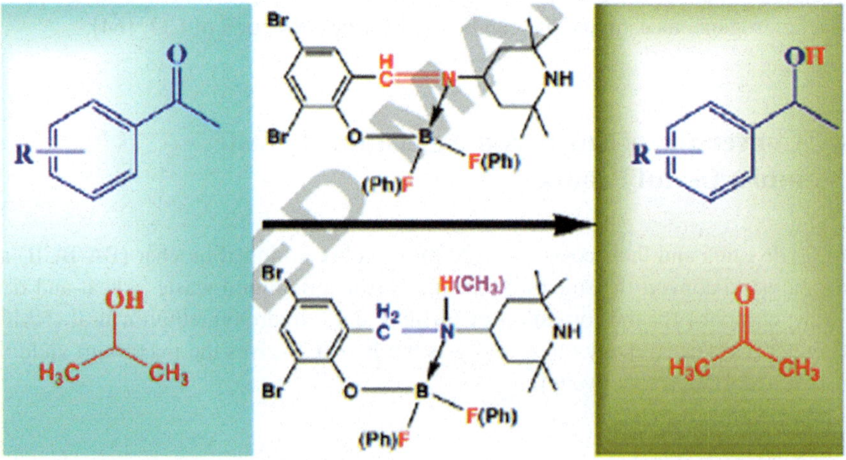

Fig. 3 $L_{(1-3)}BF_3$ or $L_{(1-3)}BPh_3$; Copyright, 2015, Spectrochimica Acta A: Molecular and Biomolecular Spectroscopy, Elsevier

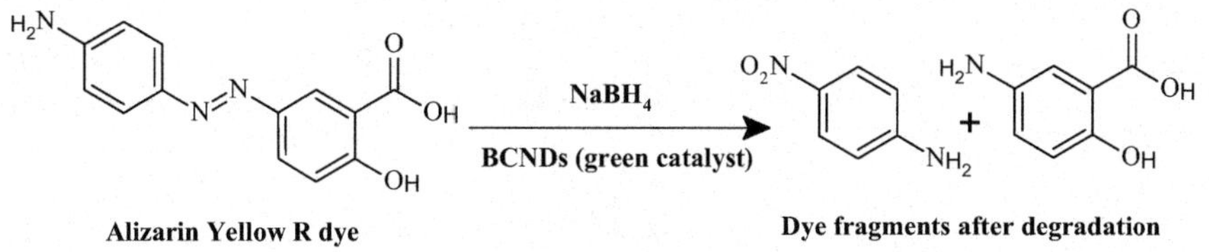

Fig. 4 BCNDs as a catalyst; Copyright, 2021, Spectrochem. Acta A: Molecular and Biomolecular Spectroscopy, Elsevier

6 Reduction of Alizarine Yellow R Dye

C. Dhanush et al. (2021) prepared boron-doped biogenic carbon nanodots (BCNDs or green catalysts) by hydrothermal method, from the extract of Syzygium cumini fruit. AYR dye (alizarine yellow R dye) was reduced by BCNDs catalysts (Fig. 4). When $NaBH_4$ and green catalysts were added with AYR dye, it showed a consistently decreasing in absorption intensity due to the breakoff of -N = N- (azo bond) and production of colorless fragments of AYR dye [66]. The reaction mechanism was based on the Langmuir–Hinshelwood mechanism. The Alizarine Y R dye and BH_4^- ions were adsorbed on the surface of the biogenic carbon nanodots catalyst during the reaction, and the movement of e^-s was takes place from BH_4^- ions to AYR dye and H_2 gas released [67].

7 Hydrogenation Reaction

S. Jungsuttiwong and their co-worker (2021) have prepared Platinum-doped boron nitride nanosheets (Pt-BNNSs) (Fig. 5). It was seen that in Pt-BV (boron vacancy site) Platinum effectively stabilized. The catalyst was used to investigate the CO_2 hydrogenation mechanism with the help of density functional theory (DFT). The catalyst shows three different reaction mechanism pathways (a) co-adsorption mechanism (b) H_2 dissociation mechanism (c) a with b pathways. On further investigation, they found that greenhouse gases (CO_2) convert into useful products (HCOOH) with the help of a Pt-BV catalyst. With the help of the reaction mechanism, they also predict the rate of reaction, kinetics, and thermodynamics [68].

Hydrogenation reaction

B.S. Kim and their co-worker (2020) have prepared the first green Gold (Au-BNNs) and silver (Ag-BNNs) nanoparticles on BNNs (boron nitride nanosheets surface) (Fig. 6) in situ, through the extract of gallnut. Au-BNNs or Ag-BNNs is used as a catalyst. When 4-nitrophenol was mixed with sodium borohydride solution in the presence of a catalyst the color of the solution changed from dark yellow to the corresponding color. Because of the Langmuir–Hinshelwood mechanism, 4-NP and BH_4^- were adsorbed on the AuNPs surface and formed HAP. HAP was further

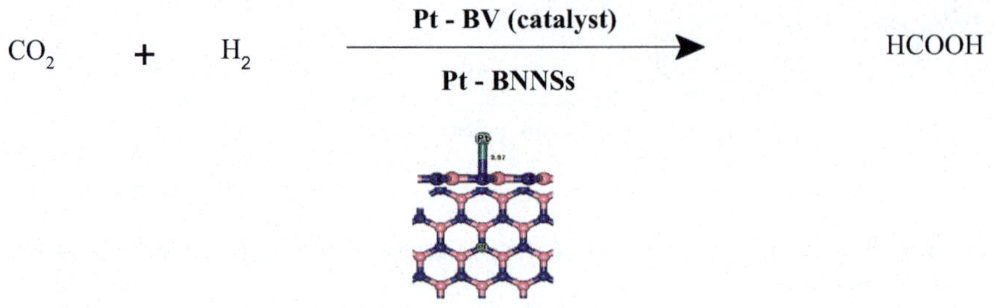

Fig. 5 Pt-BNNs as a catalyst; Copyright, 2021, Molecular Catalysis, Elsevier

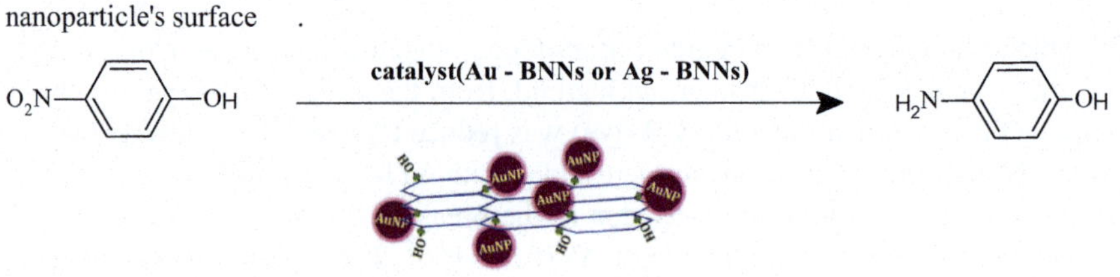

Fig. 6 Au-BNNs as a catalyst; Copyright, 2020, Journal of Cleaner Production, Elsevier

reduced to 4-aminophenol. All the reaction mechanisms was takes place on the nanoparticle's surface [69].

8 Hydrogenation Reaction

T. Wen and their co-worker (2015) have synthesized a boron-containing metal complex compound [$Cd(CH_3COO)B(im)_4$] i.e. BIF-51 (Fig. 7), by the reaction between $Cd(CH_3COO)_2{\cdot}2H_2O$ and $KB(im)_4$ in DMF and tert-butanol. After completion of the reaction, pure BIF-51 was obtained. BIF-51 was further reacted with $AgNO_3$ in DMF, which gives pure Ag@BIF-51 a catalyst, which reduces 4-nitrophenol into 4-aminophenol in the presence of KBH_4 and Ag@BIF-51 [70].

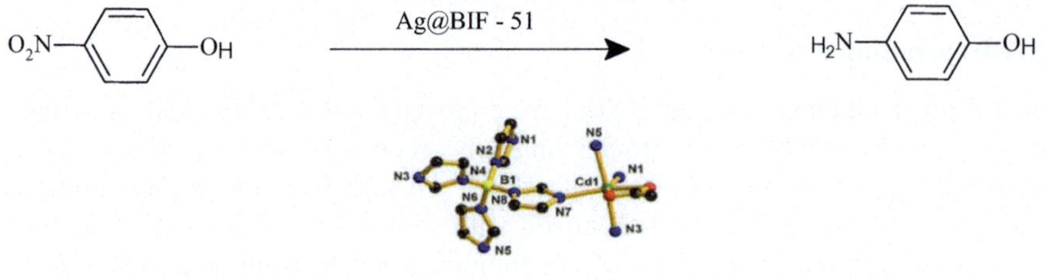

Fig. 7 Ag@BIF-51 as a catalyst; Copyright, 2015, Journal of Solid-State Chemistry, Elsevier

I.C.M.S. Santos and their co-worker (2013) designed and synthesized boron peroxotungstate complex $(C_{19}H_{34}N)_4H[BW_4O_{24}]$ or $[BW_4]$ (Fig. 7) a catalyst, by the reaction between $Na_2WO_4.2H_2O$ and H_3BO_3 in the presence of HCl (pH = 2.8). Benzyltributylammonium chloride (BTBACl) was added to the mixture. A white solid precipitate was obtained. The complex was used as a homogenous and heterogeneous catalyst.

(a) **Homogenous catalyst**

BW_4 is known as the epoxidation catalyst, because in the presence of BW_4 and H_2O_2 in acetonitrile (i) cis-cyclooctene gives corresponding epoxide (ii) geraniol gives 2,3-epoxide (iii)Linalool acetate gives furan hydroxyl ether (iv) linalyl gives 6,7-epoxide and (v) carvone obtained from (-)-carveol [71].

(b) **Heterogeneous catalyst**

BW_4@aptesSiO_2 and BW_4@MIL-101 are two different heterogeneous catalysts. When boron peroxotungstate (BW4) was added to functionalized silica (apteSiO_2) it gives BW_4@aptesSiO_2 and when BW_4 was encapsulated with MIL-101 it gives BW_4@MIL-101.

After examining the catalytic activity of heterogeneous catalyst it was observed that BW_4@MIL-101 gives better result in comparison to BW_4@aptesSiO_2 and the complete conversion of graniol take place in 3 h, while complete conversion of graniol by homogeneous catalyst takes place in 2 h. It was also observed that heterogeneous solid catalysts can be reused easily in the same condition, and used solid heterogeneous catalyst can be separated by filtration. BW_4@MIL-101 catalyst was active for three consecutive reaction cycles under the same condition, such that oxidation of graniol occurs in presence of a heterogeneous catalyst and H_2O_2 in acetonitrile [71] (Fig. 8).

9 Vanadium Redox Flow Battery

S. Liu and their co-worker (2015) have prepared a 'B' and 'N' co-doped carbon nanofibre (Fig. 9). First of all, they prepare carbon-nano-fiber (CNF) by electrospinning polyacrylonitrile followed by carbonization at 1000 °C in a furnace (Quartz furnace). Then boron and nitrogen precursors were mixed and scattered/spread on a quartz boat. The CNF and precursors were mixed equally, and finally, one more layer of precursors was added on top of the quartz boat. The mixture was heated at 1000 °C for 4 h. After cooling, it was treated with 10% of HCl for 6 h. The obtained mixture on purification gives B-CNF-1000 (B-CNF). Similarly, N-CNF was also prepared by the same procedure. B-CNF is used as a catalyst in vanadium redox flow batteries (VRFB) [72].

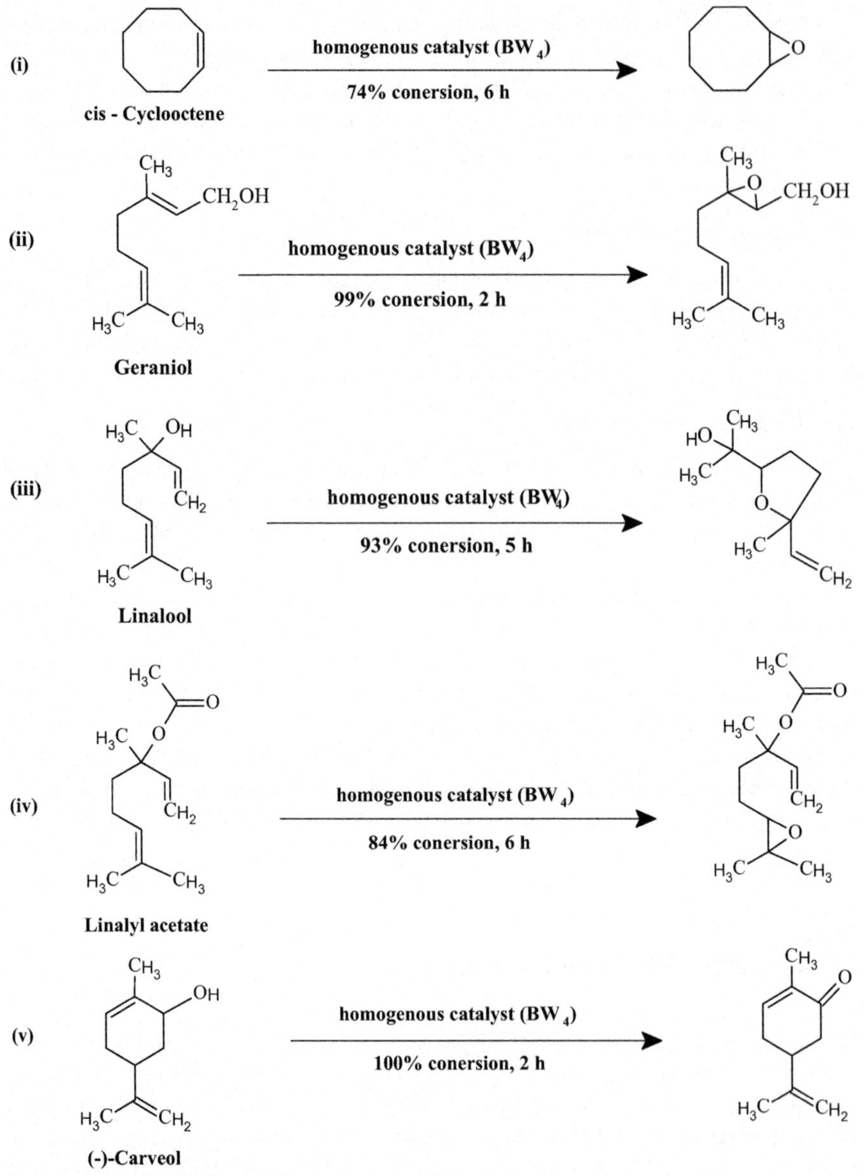

(b) Heterogeneous catalyst

Fig. 8 Schematic indicates the catalytic activity of BW4, BW4@aptesSiO_2 and BW_4@MIL-101; Copyright, 2013, Catalysis Today, Elsevier

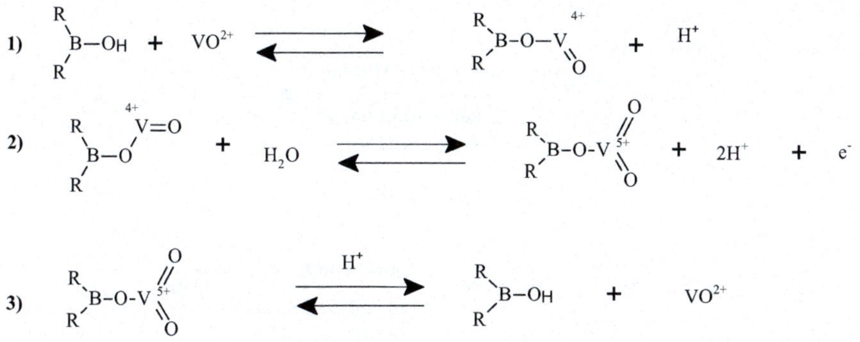

Fig. 9 Schematic structure of B-CNF catalyst for redox reaction of VO^{2+}/VO_2^{+}; Copyright, 2015, Electrochemica Acta, Elsevier

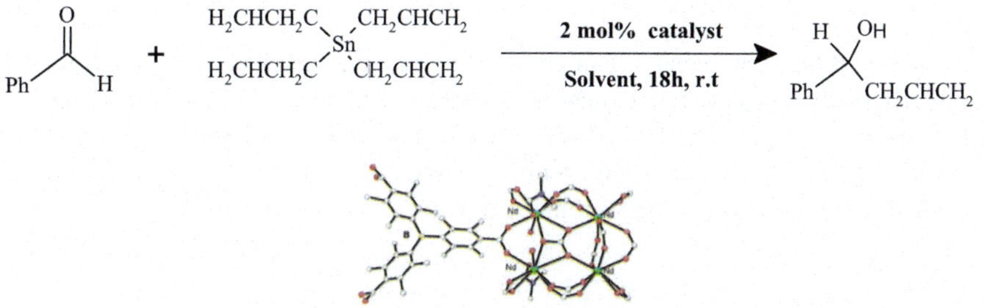

Fig. 10 Schematic representation of allylation of benzaldehyde and BMOMs catalyst; *copyright 2013, Inorganic Chemistry, American Chemical Society*

10 Allylation of Benzaldehyde in the Presence of a Catalyst

Y. Cui et al. (2013) designed and synthesis boron-based metal (Nd and La) organic materials (BMOMs) (Fig. 10). which is used as a heterogeneous catalyst, when benzaldehyde and tetra-allyltin was reacts in the presence of BMOMs it gives 99% yield. This catalyst can be reused easily for three times more without losing any significant catalytic activity [73].

11 Examination of Catalyst for Diels–Alder Reaction and Strecker-Type Reaction

The catalyst was further used to examine the Strecker-Type reaction and Diels–Alder reaction (Fig. 11), which showed a satisfactory conversion in the product. Diels–Alder conversion occurred when 3-acryloyl-1,3-oxazolidin-2-one and cyclopentadiene reacted in the presence of the 2 mol% of BMOMs catalyst and 20 mol% SDS

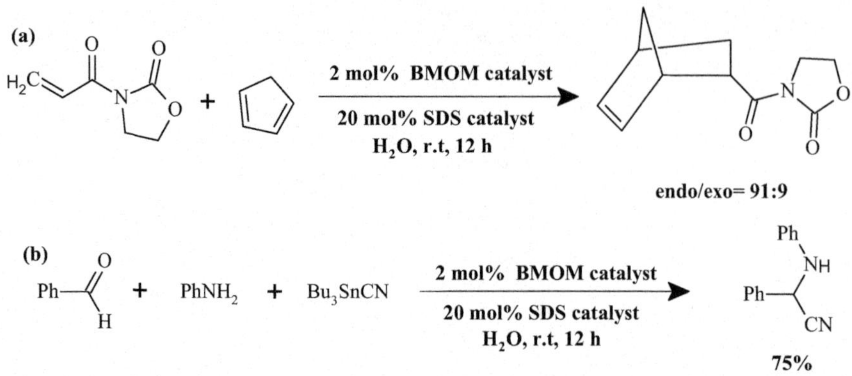

Fig. 11 Schematic indicates **a** Diels–Alder reaction, **b** Strecker-Type reaction; copyright 2013, Inorganic Chemistry, American Chemical Society

catalyst at r.t for 12 h in water, that gives the corresponding product. While Strecker-type reaction occurred when benzaldehyde reacts with aniline, tributyltin cyanide in the presence of 2 mol% of BMOMs catalyst and 20 mol% SDS catalyst at r.t for 12 h in water, gives 75% yields α-amino nitrile derivative. In both, the cases yield of the product was much higher than another catalyst [73].

12 Synthesis of Benzimidazoles

P.J. Dyson and their co-worker (2019) synthesized a boron-based metal (Eu) organic material catalyst (Eu/SION-105) (Fig. 12). Which was based on the water-stable metal–organic framework. The catalyst has application in Frustrated Lewis Pair (FLP) chemistry. Which is used as a heterogeneous catalyst for the synthesis of benzimidazoles with 99% yield, when o-phenylenediamine (*o*-diamines) was reacted with CO_2 in the presence of silane as a reducing agent. This catalyst can be used five times without losing its catalytic activity, after using of catalyst simply washed with methyl alcohol. All the catalytic activity occurs on the surface of the SION-105 catalyst [74].

13 Synthesis of CH_4

P. Dyson and et.al (2020) have designed and synthesized a Ruthenium Nanoparticle metal–organic framework catalyst (Ru/SION-105). The catalyst was used for the synthesis of CH_4 through indirect reduction of CO_2 or methanation of CO_2 (Fig. 13). This process can be repeated many times. In this process when CO_2 reacts with amino

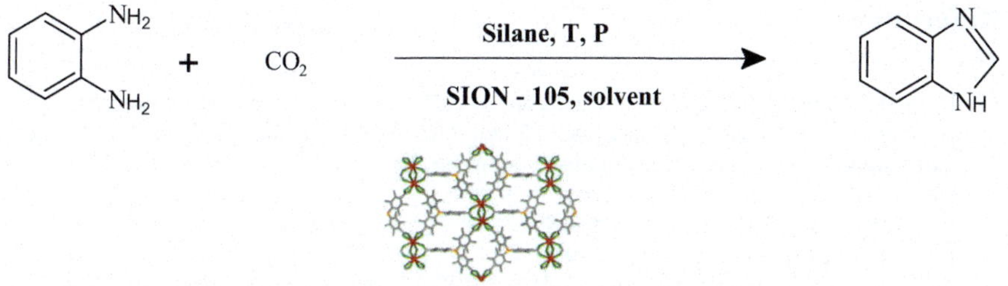

Fig. 12 Schematic diagram of reaction of aromatic o-diamines with CO_2 and SION-105 catalyst; copyright (2019), Angewandte Chemie International Edition

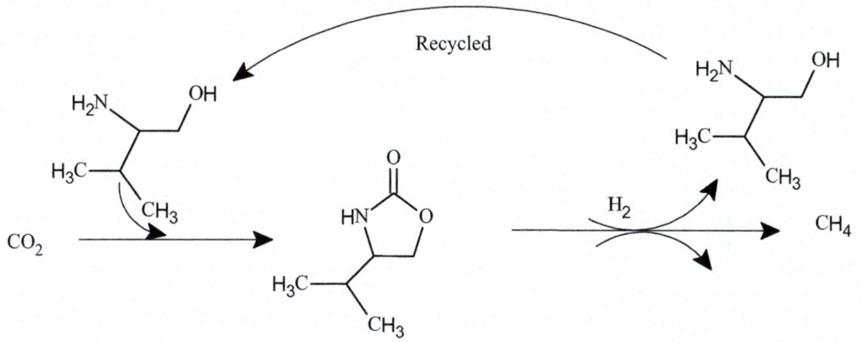

Fig. 13 Schematic indicates the conversion of CO_2 into CH_4; Copyright (2020), Angewandte Chemie International Edition

alcohols in the presence of a catalyst, it transforms into oxazolidinones, which on further reaction with H_2 changed into methane [75].

14 Conclusion

We have provided some insights on the heterocyclic metal complexes including one broan atom have been described by the different ligands such as benzothiazole, hydroxy quinoline, Different types of Schiff bases, Pyrroyl, Benzodithiaborol, and Benzo-1,3,2-oxothiaborolane, etc. Further boron metal complexes have been reported as catalytic applications of various organic reactions such as the Conversion of sugar and agricultural waste into a useful product, reduction of ketone, hydrogenation, alkylation, Diels–Alder reaction and Strecker-type reaction.

Acknowledgements The authors are highly thankful to Maulana Azad National Urdu University, Hyderabad 500032, Telangana, India for providing infrastructure facilities the clock internet facility, and other respected scientist researchers who have provided the Ph.D. of the required articles during writing this chapter.

References

1. Davy H (1809) An account of some new analytical researches on the nature of certain bodies, particularly the alkalies, phosphorus, sulphur, carbonaceous matter, and the acids hitherto undecomposed: with some general observations on chemical theory. Philos Trans R Soc Lond 99:39–104. https://doi.org/10.1098/rstl.1809.0005
2. Gay Lussac JL, Thenard LJ (1808) Sur la décomposition et la recomposition de l'acide boracique. Annales de chimie. 68:169–174
3. Holleman AF, Wiberg E, Wiberg N (1985) "Bor". Lehrbuch der Anorganischen Chemie (in German) (91–100 ed.). Walter de Gruyter 814–864. ISBN 978-3-11-007511-3
4. Kabu M, Uyarlar C, Żarczyńska K, Milewska W, Sobiech P (2015) The role of boron in animal health. J Elem 20(2):535–541. https://doi.org/10.5601/jelem.2014.19.3.706
5. Herring HW (1966) Selected mechanical and physical properties of boron filaments (PDF). NASA. Archived (PDF) from the original on 22 February 2014. Retrieved 20 September 2008
6. Kostick SD (2006) Mineral yearbook: Boron (PDF). United States Geological Survey. Archived (PDF) from the original on 20 September 2008
7. Pfaender HG (1996) Schott guide to glass, 2nd edn. Springer. p 122. ISBN 978-0-412-62060-7
8. Wisnyi LG, Taylor KL (1959) Fabrication and evaluation of Urania-Alumina fuel elements and boron carbide burnable poison elements. Archived 27 July 2020 at the wayback machine. ASTM Special Technical Publication No. 276: Materials in Nuclear Applications, Committee E-10 Staff, American Society for Testing Materials
9. Gogotsi YG, Andrievski RA (1999) Materials science of carbides, nitrides, and borides. Springer, p 270. ISBN 978-0-7923-5707-0
10. Hammond CR (2004) The elements. In Handbook of chemistry and physics, 81st edn. CRC press. ISBN 978-0-8493-0485-9
11. Klotz JH, Moss JI, Zhao R, Jr. Davis LR, Patterson RS (1994) Oral toxicity of boric acid and other boron compounds to immature cat fleas (Siphonaptera: Pulicidae). J Econ Entomol, 87(6):1534–1536.https://doi.org/10.1093/jee/87.6.1534.PMID7836612
12. May GS, Spanos CJ (2006) Fundamentals of semiconductor manufacturing and process control. Wiley, pp 51–54. ISBN 978-0-471-78406-7
13. Campbell P (1996) Permanent magnet materials and their application. Cambridge University Press, p 45. ISBN 978-0-521-56688-9
14. Pastina B, Isabey J, Hickel B (1999) The influence of water chemistry on the radiolysis of the primary coolant water in pressurized water reactors. J Nucl Mater 264(3):309–318. Bibcode:1999JNuM..264..309P. https://doi.org/10.1016/S0022-3115(98)00494-2. ISSN 0022-3115
15. "Boric acid". chemicalland21.com. Archived from the original on 3 June 2009. Retrieved 28 May 2009
16. Lin LX, Zheng Y, Wei KM (2007) Facile synthesis of hexagonal boron nitride fibers and flowers. Mater Lett 61(8–9):1735–1737
17. Chen Y, Chadderton LT, Gerald JF, Williams JS (1999) A solid-state process for the formation of boron nitride nanotubes. Appl Phys Lett 74(20):2960–2962
18. Chen Y, Gerald JF, Williams JS, Bulcock S (1999) Synthesis of boron nitride nanotubes at low temperatures using reactive ball milling. Chem Phys Lett 299(3–4):260–264
19. Lourie OR, Jones CR, Bartlett BM, Gibbons PC, Ruoff RS, Buhro WE (2000) CVD growth of boron nitride nanotubes. Chem Mater 12(7):1808–1810
20. Mickelson W, Aloni S, Han WQ, Cumings J, Zettl A (2003) Packing C60 in boron nitride nanotubes. Science 300(5618):467–469
21. Oku T, Kuno M, Kitahara H, Narita I (2001) Formation, atomic structures and properties of boron nitride and carbon nanocage fullerene materials. Int J Inorg mater 3(7):597–612
22. Deepak FL, Vinod CP, Mukhopadhyay K, Govindaraj A, Rao CNR (2002) Boron nitride nanotubes and nanowires. Chem Phys Lett 353(5–6):345–352
23. Morell G, Nocua JE, Piazza F, Weiner BR (2009) High-yield synthesis of stoichiometric boron nitride nanostructures. J Nanomater 2009, Article ID 429360

24. Zhi C, Bando Y, Tang C, Golberg D, Xie R, Sekiguchi T (2005) Large-scale fabrication of boron nitride nanohorn. Appl Phys Lett 87(6), Article ID 063107
25. Zhi C, Bando Y, Tang C, Kuwahara H, Golberg D (2009) Large-scale fabrication of boron nitride nanosheets and their utilization in polymeric composites with improved thermal and mechanical properties. Adv Mater 21(28):2889–2893
26. Essafti A, Efecto de la temperatura de deposición en las características estructurales y ópticas de películas delgadas de nitruro de boro obtenidas por CVD. Boletín de la Sociedad Española de Cerámica y Vidrio, \46(3):127–1
27. Shen BX, Qian Y (2018) Building rhodamine-BODIPY fluorescent platform using Click reaction: Naked-eye visible and Multi-channel Chemodosimeter for detection of Fe 3+ and Hg 2+," Sensors and Actuators B: Chemical
28. Salih P, Ömer E, Çiğdem Y (2019) Synthesis and structural identification of boron based schiff compounds with ishikawa endometrial cancer and antioxidant activity. J Mol Struct. https://doi.org/10.1016/j.molstruc.2019.03.061
29. Salih P (2019) Synthesis and characterization of di-Schiff based boronic structures: therapeutic investigation against cancer and implementation for antioxidant. J Mol Struct. https://doi.org/10.1016/j.molstruc.2019.05.133
30. Kilic A, Alcay F, Aydemir M, Durgun M, Keles A, Baysal A (2015) Synthesis, spectroscopic and catalytic properties of some new boron hybrid molecule derivatives by BF2 and BPh2 chelation. Spectrochim Acta A Mol Biomol Spectrosc 142:62–72. https://doi.org/10.1016/j.saa.2015.01.094
31. Kilic A, Kayan C, Aydemir M, Durap F, Durgun M, Baysal A, Tas E, Gümgüm B (2011) Synthesis of new boron complexes: application to transfer hydrogenation of acetophenone derivatives. Appl Organomet Chem 25(5):390e394
32. Chen CT (2004) Evolution of red organic light-emitting diodes: materials and devices. Chem Mater 16(23):4389e4400
33. Kraft A, Grimsdale AC, Holmes AB (1998) Electroluminescent conjugated polymersdseeing polymers in a new light. Angew Chem Int Ed 37(4):402e428
34. Entwistle CD, Marder TB (2004) Applications of three-coordinate organoboron compounds and polymers in optoelectronics. Chem Mater 16(23):4574e4585
35. Wang D, Fan F, Gao X, Wang B, Sun S, Peng X (2009) Carboxyl BODIPY dyes from dicarboxylic anhydrides: one-pot preparation, spectral properties, photostability, and biolabeling. J Org Chem 74(20):7675e7683
36. Guliyev R, Buyukcakir O, Sozmen F, Bozdemir OA (2009) Cyanide sensing via metal ion removal from a fluorogenic BODIPY complex. Tetrahedron Lett 50(36):5139e5141
37. Crandall LA, Rhoda HM, Nemykin VN, Ziegler CJ (2016) Boron templated synthesis of a BODIPY analogue from a phthalocyanine precursor. New J Chem 40(7):5675–5678. https://doi.org/10.1039/c6nj00085a
38. Crossley DL, Cade A, Clark ER, Escande A, Humphries MJ, King SM, Vitorica-Yrezabal I, Ingleson MJ, Turner ML (2015) Enhancing electron affinity and tuning band gap in donor-acceptor organic semiconductors by benzothiadiazole directed C–H borylation. Chem Sci 6(9):5144–5151. https://doi.org/10.1039/c5sc01800e
39. Sadu VS, Bin HR, Lee DM, Lee KI (2017) "One-pot synthesis of four coordinate boron (III) complexes by the ligand-promoted organic group migration between boronic acids. Sci Rep 7:242. https://doi.org/10.1038/s41598-017-00236-2
40. Brothers PJ (2011) Boron complexes of pyrrolyl ligands. Inorg Chem 50(24):12374–12386. https://doi.org/10.1021/ic2011919
41. López MMC, Pérez VMJ, Navarro RC, Rodríguez MI, Dias HVR, Reyes AC, Flores BMM (2017) Synthesis, characterization, photophysical properties of new fluorescent boron Schiff bases (BOSCHIBAs) and their application as cytoplasm staining dyes in vitro. J Organomet Chem 852:64–73. https://doi.org/10.1016/j.jorganchem.2017.10.003
42. Wakamiya A, Taniguchi T, Yamaguchi S (2006) Angew Chem Int Ed 45:3170–3173; Angew Chem 118:3242–3245

43. Hecht R, Kade J, Schmidt D, Nowak-Król A (2017) n-channel organic semiconductors derived from air-stable four-coordinate boron complexes of substituted thienylthiazoles. Chem Eur J 23(48):11620–11628.https://doi.org/10.1002/chem.201701922
44. Dou C, Long X, Ding Z, Xie Z, Liu L, Wang L (2015) An electron-deficient building block based on the B←N unit: an electron acceptor for all-polymer solar cells. Angew Chem Int Ed 55(4):1436–1440. https://doi.org/10.1002/anie.201508482
45. Hadebe KN, et al (2013) The synthesis and application of heteroatom borolanes containing nitrogen, oxygen, and sulphur" research space. https://researchspace.ukzn.ac.za/handle/10413/8888
46. Tas E, Kilic A, Konak N, Yilmaz I (2008) The sterically hindered salicylaldimine ligands with their copper (II) metal complexes: synthesis, spectroscopy, electrochemical and thin-layer spectroelectrochemical features. Polyhedron 27(3):1024–1032. https://doi.org/10.1016/j.poly.2007.11.038
47. Kilic A, Ozbahceci O, Durgun M, Aydemir M (2017) Different hemi-salen/salan ligand containing binuclear boron-fluoride complexes: synthesis, spectroscopy, fluorescence properties, and catalysis. Polycycl Aromat Compd 1–18. https://doi.org/10.1080/10406638.2017.1314973
48. Bukvetskii BV, Fedorenko EV, Mirochnik AG (2011) Crystal structure and luminescent properties of 2,2-difluoro-4-(9H-fluorene-2-yl)-6-methyl-1,3,2-dioxaborine. J Struct Chem 52(1):221–225
49. Liu B, Novikova N, Simpson MC, Timmer MSM, Stocker BL, Söhnel T, Brothers PJ (2016) Lighting up sugars: fluorescent BODIPY–gluco-furanose and –septanose conjugates linked by direct B-O–C bonds. Org Biomol Chem 14(23):5205–5209. https://doi.org/10.1039/c6ob00726k
50. Umezawa K, Matsui A, Nakamura Y, Citterio D, Suzuki K (2008) Bright, color-tunable fluorescent dyes in the vis/nir region: establishment of new "Tailor-Made" multicolor fluorophores based on borondipyrromethene. Chem Eur J 15(5):1096–1106. https://doi.org/10.1002/chem.200801906
51. Lerrick RI, Winstanley TPL, Haggerty K, Wills C, Clegg W, Harrington RW, Hall MJ (2014) Axially chiral BODIPYs. Chem Commun 50(36):4714–4716. https://doi.org/10.1039/c4cc00851k
52. Sirbu D, Benniston AC, Harriman A (2017) One-pot synthesis of a mono-O, B, N-strapped BODIPY derivative displaying bright fluorescence in the solid state. Org Lett 19(7):1626–1629. https://doi.org/10.1021/acs.orglett.7b00435
53. Courtis AM, Santos SA, Guan Y, Hendricks JA, Ghosh B, Szantai-Kis DM, Mazitschek R (2014) Monoalkoxy BODIPYs—a fluorophore class for bioimaging. Bioconjug Chem 25(6):1043–1051.https://doi.org/10.1021/bc400575w
54. Bauman GW, Jastrzębski L, Kurach P, Luliński S, Serwatowski J, Woźniak K (2012) Synthesis of functionalized diarylborinic 8-oxyquinolates via bimetallic boron–lithium intermediates. J Organomet Chem 711:1–9. https://doi.org/10.1016/j.jorganchem.2012.03.002
55. Jarzembska KN, Kamiński R, Durka K, Kubsik M, Nawara K, Witkowska E, Woźniak K (2017) New class of easily synthesizable and modifiable organic materials for applications in luminescent devices. Dyes and Pigm 138:267–277. https://doi.org/10.1016/j.dyepig.2016.11.039
56. Davis BL, Dixon DA, Garner EB, Gordon JC, Matus MH, Scott B, Stephens FH (2009) Efficient regeneration of partially spent ammonia borane fuel. Angew Chem Int Ed 48(37):6812–6816. https://doi.org/10.1002/anie.200900680
57. Ji L, Zhou H (2021) Recent developments in the synthesis of bioactive boron-containing compounds. Tetrahedron Lett 153411. https://doi.org/10.1016/j.tetlet.2021.153411
58. Baykal A, Kizilkyalli M, Toprak M, Knicp R (2001) Turk J Chem 25(4):425–432
59. Sun W, Meng Y, Fu Q, Wang F, Wang G, Gao W, Lu F (2016) High-yield production of boron nitride nanosheets and its uses as a catalyst support for hydrogenation of nitroaromatics. ACS Appl Mater Interfaces 8(15):9881–9888. https://doi.org/10.1021/acsami.6b01008

60. Avcioglu S, Kaya F, Kaya C (2019) Non-catalytic synthesis of boron carbide (B4C) nanostructures with various morphologies by sol-gel process. Mater Lett. https://doi.org/10.1016/j.matlet.2019.04.056
61. Cao LM, Hahn K, Scheu C, Ruhle M, Wang YQ, Zhang Z, et al. (2002) Template-catalyst-free growth of highly ordered boron nanowire arrays. Appl Phys Lett 80(22):4226–4228
62. Ferdous D, Dalai AK, Adjaye J (2004) A series of NiMo/Al2O3 catalysts containing boron and phosphorus. Appl Catal A 260(2):137–151. https://doi.org/10.1016/j.apcata.2003.10.010
63. Szeto KC, Sahyoun W, Merle N, Castelbou JL, Popoff N, Lefebvre F, Taoufik M (2016) Development of silica-supported frustrated Lewis pairs: highly active transition metal-free catalysts for the Z-selective reduction of alkynes. Catal Sci Technol 6(3):882–889. https://doi.org/10.1039/c5cy01372k
64. Bontemps S (2016) Boron-mediated activation of carbon dioxide. Coord Chem Rev 308:117–130. https://doi.org/10.1016/j.ccr.2015.06.003
65. Ofrasio BIG, de Luna MDG, Chen YC, Abarca RRM, Dong CD, Chang KL (2020) Catalytic conversion of sugars and biomass to furanic biofuel precursors by boron-doped biochar in ionic liquid. Bioresour Technol 100515. https://doi.org/10.1016/j.biteb.2020.100515
66. Zhang Y, He P, Jia L, Li C, Liu H, Wang S, Zhou S, Dong F (2019) Ti/PbO2-Sm2O3 composite based electrode for highly efficient electrocatalytic degradation of alizarin yellow R. J Colloid Interface Sci 533:750–761. https://doi.org/10.1016/j.jcis.2018.09.003
67. Dhanush C, Sethuraman MG (2021) Independent hydrothermal synthesis of the undoped, nitrogen, boron and sulphur doped biogenic carbon nanodots and their potential application in the catalytic chemo-reduction of Alizarine yellow R azo dye. Spectrochim. Acta A Mol Biomol. Spectrosc 260:119920. https://doi.org/10.1016/j.saa.2021.119920
68. Injongkol Y, Intayot R, Yodsin N, Montoya A, Jungsuttiwong S (2021) Mechanistic insight into catalytic carbon dioxide hydrogenation to formic acid over Pt-doped boron nitride nanosheets. Mol Catal 510:111675. https://doi.org/10.1016/j.mcat.2021.111675
69. Deshmukh AR, Aloui H, Kim BS (2020) In situ growth of gold and silver nanoparticles onto Phyto-functionalized boron nitride nanosheets: catalytic, peroxidase mimicking, and antimicrobial activity. J Clean Prod 122339. https://doi.org/10.1016/j.jclepro.2020.122339
70. Liu L, Wen T, Chen S, Zhang J (2015) A zeolitic Cd(II) boron imidazolate framework with sensing and catalytic properties. J Solid State Chem 231:185–189. https://doi.org/10.1016/j.jssc.2015.08.033
71. Santos ICMS, Balula SS, Simões MMQ, \Silva LC, Neves MGPMS, de Castro B, Cavaleiro JAS (2013) Catalytic performance of a boron peroxotungstate complex under homogeneous and heterogeneous conditions. Catal Today 203:87–94. https://doi.org/10.1016/j.cattod.2012.03.071
72. Shi L, Liu S, He Z, Yuan H, Shen J (2015) Synthesis of boron and nitrogen co-doped carbon nanofiber as efficient metal-free electrocatalyst for the VO 2+ /VO 2 + Redox Reaction. Electrochim Acta 178:748–757. https://doi.org/10.1016/j.electacta.2015.08.026
73. Liu Y, Mo K, Cui Y (2013) Porous and robust lanthanide metal-organoboron frameworks as water tolerant lewis acid catalysts. Inorg Chem 52(18):10286–10291. https://doi.org/10.1021/ic400598x
74. Shyshkanov S, Nguyen T, Ebrahim FM, Stylianou K, Dyson PJ (2019) In situ formation of frustrated lewis pairs in a water-tolerant metal-organic framework for the transformation of CO2. Angew Chem Int Ed. https://doi.org/10.1002/ange.201901171
75. Dyson PJ, Cui X, Shyshkanov S, Nguyen TN, Chidambaram A, Fei Z, Stylianou KC (2020) CO2 Methanation via amino-alcohol relay molecules employing a ruthenium nanoparticle-metal-organic framework catalyst. Angew Chem Int Ed. https://doi.org/10.1002/anie.202004618

Thermally Conductive Hexagonal Boron Nitride-Polyethylene Nanocomposites: Effect of Processing Method and Filler Size, Exfoliation, and Alignment on Thermal Conductivity

Mehamed Ali and Ahmed Abdala

Abstract Applications of polymers in thermal management have recently gained significant interest. Rapid heat transfer demand in thermal management applications requires enhancing the low thermal conductivity of polymers. We developed polyethylene (PE)-hexagonal boron nitride (hBN) composites with hBN loading between 5 and 40 vol%. The impacts of the composite fabrication method, the PE type, and hBN size, exfoliation, and alignment on the composite morphology, thermal conductivity, thermal stability, and mechanical properties were thoroughly investigated. Our results indicate increasing the hBN loading increases κ_c and enhance the mechanical properties and the thermal stability of the composites. Moreover, dry ball milling of hBN and PE yields composites with higher κ_c than melt extrusion. The processing-induced hBN alignment significantly affects κ_c. Melt extrusion followed by compression molding produces composites with hBN aligned parallel to the sheet surface at the skin layers (~fixed 250 μm each) and randomly aligned at the core of the sheet. This processing-induced hBN alignment led to κ_c dependence on the sheet thickness as the core layer has higher conductivity than the skin layers. By controlling the alignment of hBN in LLDPE, κ_c was greatly enhanced, reaching up to ~600 increase in the κ_c of the unaligned hBN composites. Finally, the exfoliation of hBN proved effective in increasing κ_c and improving the composites' mechanical properties and thermal stability but led to significant processing challenges due to the very high melt viscosity.

Keywords Polyethylene · Thermal conductivity · Nanocomposite · hBN · hBN exfoliation · Ball milling

---

M. Ali · A. Abdala (✉)
Chemical Engineering Program, Chemical Engineering, Texas A&M University at Qatar, Doha, Qatar
e-mail: ahmed.abdala@qatar.tamu.edu

M. Jawaid and A. Khan (eds.), *Multifunctional Boron-Nitride Composites*, Composites Science and Technology, https://doi.org/10.1007/978-981-99-2866-8_3

1 Introduction

Owing to their exceptional attributes, such as lightweight, corrosion resistance, ease of fabrication, and low cost, polymers are desirable for many applications [1, 2]. Recently, the applications of polymers in thermal management, such as electronics, heat exchangers, appliances, and machinery, has remarkably increased [2–5]. However, polymers are generally insulators with phonons as the dominant heat transporters, with thermal conductivity (κ) in the range of 0.1–0.5 W m^{-1} K^{-1} [6]. Weak chain van der Waals bonds and complex morphology contribute immensely to the phonon scattering, hence their low thermal conductivity [4, 7–10]. Nevertheless, κ of polymers is frequently enhanced by the incorporation of thermally conductive inorganic fillers, such as metals [11], graphite [12], carbon fiber (CF) [13], carbon nanotubes (CNT) [14], graphene [15], aluminum nitride [16], silicon nitride [17], silicon carbide [18], and hexagonal boron nitride (hBN) [19]. Fillers improve the thermal conductivity of the polymer composites (κ_c) by forming thermally conductive networks or pathways, which act as bridges for heat transfer [6, 20].

Recently, hBN and its 2D nanosheets isoelectronic analogs to graphite and graphene, respectively, have been frequently used [21]. hBN and its 2D nanosheets possess superb mechanical properties, electrical insulating properties with bandgap of ~5.5 eV, and outstanding thermally and chemically stability [1, 22–25]. The combination of these properties plays a major role in their applications, such as dielectric substrates for electronic devices [26, 27], ceramics [28, 29], nanocomposites [30], coating [31], lubricants [32], cosmetics [33], catalyst [34], medicine [35], and automotive [36]. Moreover, the through-plane and in-plane κ of hBN is ~2–30 and 300–600 W m^{-1} K^{-1}, respectively, making hBN suitable for thermally conductive polymer composites [6, 37].

Generally, the thermal conductivity of composites is a function of filler content or loading, intrinsic thermal conductivity, type, particle size, particle shape, orientation, aspect ratio, and treatment [2, 15, 19, 30, 38–41]. For instance, Jiang et al. [38] reported 60%, 203%, and 505% improvements in the thermal conductivities of epoxy upon the addition of 3, 5, and 15 vol. % of surface-treated hBN. Similarly, Song et al. [42] prepared hBN/polyimide composites with κ_c of 7 W m^{-1} K^{-1} at 60 wt. % hBN concentration. Yet, there are practical limitations to the amount of filler that can be incorporated. For example, high loading might lead to filler agglomeration, poor mechanical performance, difficulty in fabrication, and high cost [2, 5, 20].

Similarly, the fillers' intrinsic conductivity (κ_f) is indispensable for improving the conductivity of the composites' κ_c. For example, κ_f of graphene is 5000 W m^{-1} K^{-1} [43], and that of hBN is 30–300 W m^{-1} K^{-1} [6, 37], which suggests a significant difference in κ_c of their respective composites. However, Bigg observed insignificant improvement in the composites' κ_c as the ratio between the filler κ_f and the polymer conductivity (κ) exceeds 100 [3]. Also, composites loaded with large fillers have fewer polymer-filler interfaces than those filled with smaller size fillers, leading to lower interfacial thermal resistance and higher 'κ' [2]. Shin et al. [19] observed that hBN of larger particle size to be more effective in improving κ_c of high-density polyethylene (HDPE) composites. Compared to hBN with other dimensionality, 2D BN nanosheets (hBNN) have higher thermal conductivity and better dispersibility making them more suitable in nanocomposites applications [44]. Lin et al. [30] indicated that κ_c of epoxy nanocomposite containing 5 wt. % hBNN was enhanced by 113%, whereas 28 wt. % of 3D hBN was required to achieve the same enhancement. Another critical factor is filler orientation, which is very common in non-spherical fillers such as platelet BN, fibers, carbon nanotubes, graphene, and graphite flakes, leading to anisotropic κ_c [2]. Sun et al. [39] reported a high κ_c of 3.09 W m^{-1} K^{-1} at loading of 18 vol.% of aligned hBN in polycarbonate matrix, which was much higher than κ_c of the composite with randomly aligned BN.

In this work, PE composites and nanocomposites based on hBN filler were prepared and optimized by different processing techniques. The effect of hBN loading, lateral size, number of layers, polymer type (PE type), and composites fabrication methods on the thermal conductivity was investigated. Moreover, the thermal stability and tensile properties of these composites were studied.

2 Experimental Procedure

2.1 Materials

Hexagonal boron nitride (hBN) powder with 6 and 35 μm average particle sizes and 99.0% purity were acquired from Asbury Carbon. Low-density polyethylene (LDPE) pellets and linear low-density polyethylene (LLDPE) powder with densities of 0.923 g/ cm^3 and 0.916 g/cm^3 and melting points of ~108 °C and ~122 °C, respectively, were kindly supplied by Qatar Petrochemical Company (QAPCO). High-density polyethylene (HDPE) pellets with a density of 0.938 g/cm^3 and a melting point of ~134 °C were kindly provided by Qatar Chemical Company (Q-Chem).

2.2 hBN Exfoliation Treatment

hBN was mechanically milled using Retsch CryoMill (Ball Mill). In each run, 3 g of 80 wt. % bulk hBN + 20 wt. % LLDPE and 8 mm diameter steel balls (~2 g/ball) were loaded into a steel vial. The ball to powder ratio, the milling frequency, and the milling time were 10:1, 30 Hz, and 90 min, respectively.

2.3 Nanocomposite Fabrication

hBN/PE composites and milled hBN/PE nanocomposites were prepared by melt blending (BM) and ball milling (BM) methods. MB was carried using Xplore Twin Screw 15 ml Micro-Compounder. In each batch, 9 g of the composites (PE and hBN or M-hBN powder) were premixed and fed to the extruder operating at 210 °C, 100 rpm speed, and 10 min mixing time. The hBN/PE composites were also prepared by BM via a dry mixing method using Retsch CryoMill. Here, 3 g of the composite and 8 mm diameter steel balls (~2 g/ball) were loaded to a steel vial in each run. The ball to powder ratio, the milling frequency, and the milling time were 10:1, 30 Hz, and 45 min, respectively. Finally, the composites were pressed into sheets of various thicknesses, i.e., 0.5 and 2 mm, using Craver Bench Top Heated Manual Press (Model 4386), at 180 °C and pressure of ~10 bar for 5 min.

2.4 Characterization of hBN/PE Nanocomposites

The SEM analysis of hBN and the composites were carried using VERSA 3D (FEI). TGA was conducted using Discovery TGA (TA Instruments) under nitrogen atmosphere at a heating rate of 10 °C/min and a temperature range of 25–700 °C. The DSC was carried using Q2000 DSC (TA Instruments) at a heating and cooling rate of 10 °C/min and a temperature range of 25–200 °C. The thermal conductivity (κ_c) was determined by measuring the thermal diffusivity using laser flash (LINSEIS LFA 500) at RT. The tensile properties were measured using an MTS Insight Electromechanical testing system with a load cell of 1 kN at a strain rate of 10 mm/min.

3 Results and Discussions

3.1 Morphology of hBN and hBN/PE Composites

Figure 1 shows the SEM images of hBN6 and hBN35, respectively. The hBN particles of both particle sizes are shown as agglomerated multilayered sheets. Their sizes ranged from 5 to 10 μm (hBN6) and 10 to 50 μm (hBN35), agreeing with the manufacturer's data.

Figure 2a–f show the cross-sectional SEM images of neat PE and hBN/PE composites' fractured surfaces. The PE has a smooth surface morphology, as shown in Fig. 2a. Meanwhile, the composites, notably those prepared by the MB method, displayed two main layers with different hBN arrangements relative to the heat flow direction, i.e., a skin layer at the specimen top and bottom surfaces and a core layer at the center.

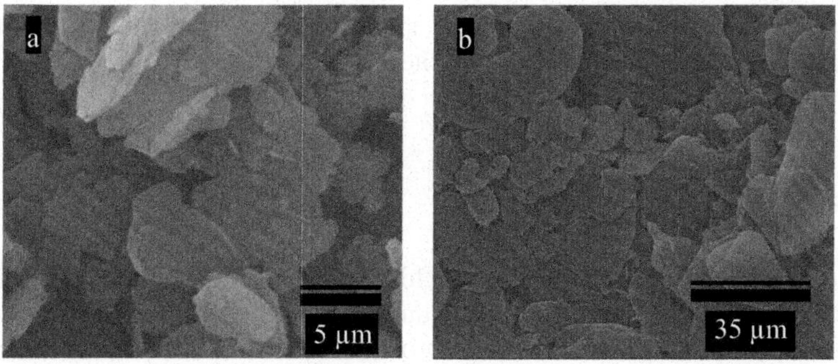

Fig. 1 SEM images of **a** hBN6 and **b** hBN35

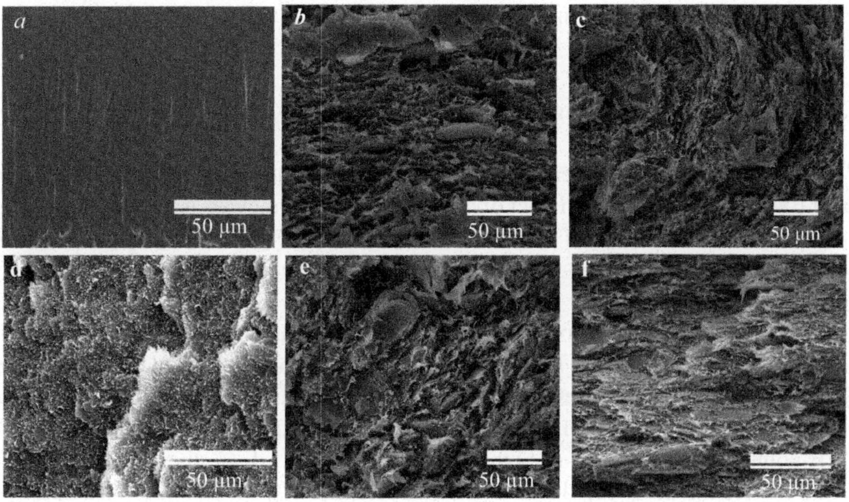

Fig. 2 Cross-sectional SEM images of fractured surfaces of **a** pure PE; **b** the skin layer of 2-mm sheet of hBN35/PE 30% MB, and the core layer of **c** 2-mm sheet of hBN35/PE 30% MB, **d** 2-mm sheet of hBN6/PE 30% MB **e** 2-mm sheet of hBN35/PE 30% BM, and **f** 0.5-mm sheet of hBN35/PE 30% MB

The skin layer showed predominately perpendicular hBN arrangement, whereas the core layer displayed mainly parallel hBN alignment. Figure 2b–d show the fractured surfaces of 2 mm thick hBN35/PE and hBN6/PE composites prepared by the MB method. The skin and core layers of hBN35/PE composites are shown in Fig. 2b, c, respectively. Similarly, hBN6/PE composites exhibited both layers. However, only the core layer was included in this work, as shown in Fig. 2d. Conversely, the 2 mm thick composite (hBN35/PE) prepared by the BM method exhibited primarily random hBN alignment (Fig. 2e). On the other hand, the 0.5 mm thick composite (hBN35/PE) prepared by the MB method, shown in Fig. 2f, mainly displayed perpendicular hBN alignment, similar to the hBN alignment exhibited in the skin layer of the 2 mm thick composite.

3.2 *Thermal Conductivity*

Generally, the thermal conductivity of hBN/PE composites increased with hBN loading because the intrinsic thermal conductivity of hBN is much higher than that of the pure PE polymers. However, this improvement in κ_c was not linear. For instance, the improvement in κ_c was insignificant at low hBN content, while at high hBN loading, due to the formation of thermal networks, there was a notable improvement in κ_c. In the following sections, the impacts of several factors on the thermal conductivity of hBN/PE composites were discussed.

3.2.1 Effect of Processing Method

To study the effect of the blending method, the composites were prepared by the MB and BM methods, and the impact of compression molding was assessed by pressing the composite into sheets of 0.5- and 2-mm thickness. Figure 3 displays the thermal conductivities of 0.5- and 2-mm thick hBN35/LLDPE composite sheets prepared by the MB and BM methods.

Regardless of the hBN content, the composites prepared by the BM method (BM-composites) demonstrated higher κ_c than MB-composites. At intermediate hBN loading, the BM-composites manifested a noticeably high κ_c as the BM composite with 22.5 vol% had κ_c of 1.15 W m^{-1} K^{-1}, compared to 0.8 W m^{-1} K^{-1} for the MB-composites. However, at high hBN loading, ~40 vol. %, there was no significant difference in κ_c for the BM and MB composites. At low hBN loading, the formation of a filler network is vital. Thus, the composites that form more effective filler thermal bridges, i.e., BM-composites, displayed higher κ_c. On the other hand, both composites formed filler networks at a very high hBN loading because of the

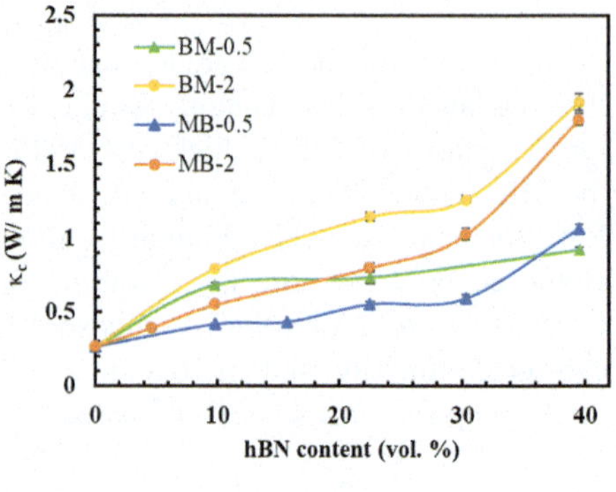

Fig. 3 Thermal conductivities hBN35/LLDPE prepared by BM and MB and pressed into 0.5- and 2-mm thick sheets as a function of hBN loading

increased packing, leading to similar κ_c. Kim et al. [45] also reported higher κ_c for hBN/polyphenylene sulfide composites prepared by a powder mixing method than those prepared by MB method. Generally, MB tends to induce hBN alignment in the melt flow direction (parallel to the sheet surface) during extrusion [46]. In contrast, the composites are mixed in a dry powder form in the BM method, resulting in composites with random filler orientation. The alignment of the anisotropic hBN sheets leads to composites with anisotropic $\kappa_{c,}$ high in the in-plane direction and lower in the orthogonal (heat flow) direction [37]. It is noteworthy to reiterate that the laser flash techniques measure the orthogonal κ_c.

To assess the morphology of these composites, the SEM analysis of their cross-sectional surface was carried, as shown in Fig. 2a–f. As mentioned earlier, the MB-composites exhibited two layers; skin layer with perpendicular hBN orientation and core layer with parallel hBN arrangement. This skin–core shell arrangement has been observed in fiber-filled composites [47–49]. The MB-composites contained two skin layers and one core layer, and the average thickness of the skin layer was ~0.4 mm. Therefore, the measured thermal conductivity is the effective κ_c of the three layers, with high κ in the core layer than the skin layer. On the contrary, the BM-composites displayed primarily random hBN arrangement (Fig. 2e), i.e., with a negligible skin layer thickness, which appeared to be caused by the compression molding, justifying their relatively improved κ_c.

Figure 3 also compares κ_c of composites of different sheet thicknesses. The 2 mm thick sheets exhibited higher κ_c than the 0.5 mm thick sheets. For example, at ~40 vol. % hBN loading κ_c of the 0.5- and 2-mm thick sheets prepared by MB were 1.07 and 1.8 W m^{-1} K^{-1}, respectively, and those fabricated by BM were 0.92 and 1.9 W m^{-1} K^{-1}. Due to its 2D sheet-like geometry, hBN aligns parallel to the surface during compression molding [39]. In the 0.5 mm thick sheets, due to the limited space, the hBN particles were forced to align perpendicular to the compression and heat flow direction, as shown in Fig. 2f.

In contrast, in the 2 mm thick sheets, the hBN particles have more freedom to take different orientations, forming skin and core layer arrangements. For instance, the total thickness of the skin layers of the 2 mm thick sheets was ~0.5–0.7 mm, which was approximately equal to the composite sheet thickness of the 0.5 mm thick sheets (0.5 mm). The alignment of hBN in the skin layers leads to a lower κ_c, because κ_f of hBN in the orthogonal direction is much higher than the parallel direction. Therefore, the 2 mm thick composites having slightly improved hBN orientations exhibited somewhat higher κ_c. It is worth noting that similar trends were evident in all composites, independent of the PE type or the filler size. The hBN orientation induced by composite sheet thickness will be further addressed and proved in detail in Sect. 3.2.3.

3.2.2 Effect of Polymer Type and hBN Particle Size

To study the effect of the polymer matrices on κ_c, three different PE polymers were used, i.e., HDPE, LDPE, and LLDPE. HDPE and LDPE were pellets, and therefore, their composites were prepared using the MB method only. However, LLDPE was available in a powder form; thus, it was fabricated using both the MB and BM methods. The effect of hBN particle size was investigated by preparing hBN6 and hBN35/PE composites via MB and BM.

Figure 4a shows the thermal conductivity of hBN3/LDPE, hBN35/LLDPE, and hBN35/HDPE composites. At low hBN loading, LDPE and LLDPE showed similar κ_c. However, at high hBN loading, LLDPE composites had slightly higher κ_c. Conversely, at all hBN loading, HDPE composites had higher κ_c than the other composites. For example, κ_c of 40 vol.% hBN composites of LDPE, LLDPE, and HDPE were 1.40, 1.80, and 2.72 W m^{-1} K^{-1}, corresponding to 445, 570, and 607% increase over κ of the neat polymers, respectively. The improved κ_c of the HDPE composites was attributed to the higher intrinsic κ of HDPE polymer because of its higher crystallinity and lower branching content, and hence lower phonon scattering. In addition, the high melt viscosity of the HDPE composites could reduce the alignment of the hBN sheets during the compression molding due to their slower diffusion, leading to higher κ_c enhancement. This will be discussed in more detail in Sect. 3.2.3.

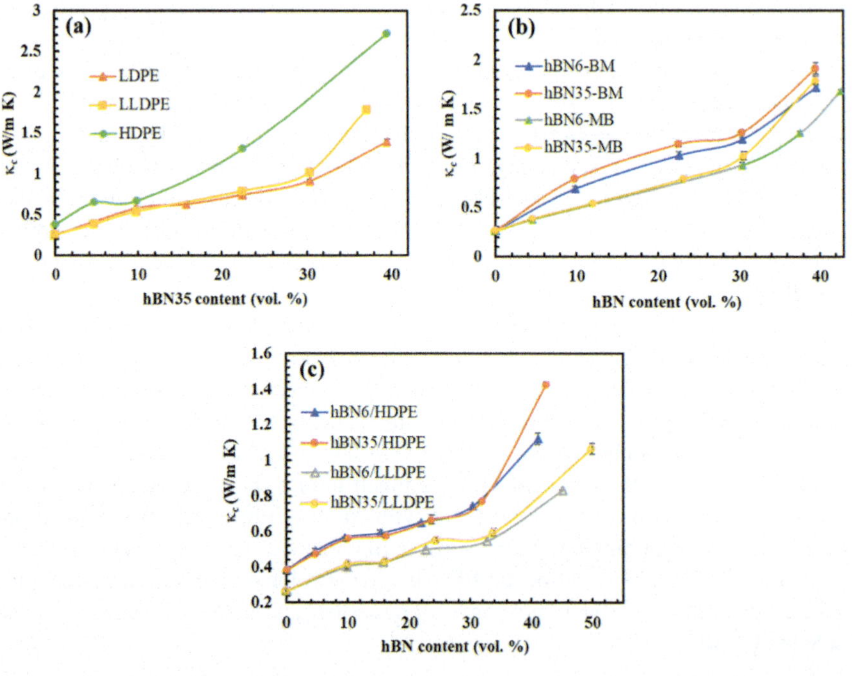

Fig. 4 **a** κ of 2 mm thick hBN35/LDPE, hBN35/LLDPE, and hBN35/HDPE composites prepared by MB, **b** κ of 2 mm thick hBN6 and hBN35/LLDPE composites prepared by BM and MB methods pressed into 2 mm thick sheets, and **c** κ of 0.5 mm hBN6/LLDPE, hBN6/HDPE, hBN35/LLDPE, and hBN35/HDPE composites prepared by MB

Figure 4b shows κ_c of hBN6/LLDPE and hBN35/LLDPE composites prepared by the BM and MB methods. Of the two-particle sizes, hBN35 filled composites showed slightly higher κ_c. As aforementioned, big fillers tend to form networks at relative ease, creating composites with reduced thermal interface resistance and enhanced κ_c [2]. The SEM images of hBN35/LLDPE and hBN6/LLDPE are shown in Fig. 2a–f. The hBN particles in the hBN35/LLDPE formed networks at a relatively lower number of hBN particles than those of the hBN6/LLPDE, thereby reducing the filler–polymer interfaces. However, the difference in their κ_c was not substantial, perhaps, because the two hBN particles size exhibited similar aspect ratios. For example, the calculated average aspect ratios of hBN6 and hBN35 were 30 and 18, respectively, justifying this result. These results were also obtained in other composites regardless of their sheet thickness, fabrication techniques, or polymer matrices, as shown in Fig. 4b, c.

3.2.3 Effect of hBN Alignment

Due to their plate-like geometry, hBN particles tend to arrange themselves during processing, such as extrusion and compression molding, resulting in composites with anisotropic κ_c. This factor, hBN alignment, was proved to be the most vital factor in improving κ_c) thoroughly studied.

First, the skin and core layer structures were studied. Here, the skin layer of the 2 mm thick composites was removed by polishing at an interval of ~0.5 mm. Figure 5a shows κ_c of 39 vol. % hBN35/HDPE and hBN35/LLDPE composites at each removed skin layer thickness. In addition, the skin removal is illustrated by the colored shapes; black being the outermost layer (skin), and red being the innermost layer (core). As expected, κ_c improved as the removed skin layer thickness increased. Notably, as the removed skin layer increased from 0 to 0.74 mm, κ_c of the 39 vol.% hBN35/HDPE rose from 2.72 to 3.44 W m^{-1} K^{-1}, however, further removing the skin layer resulted in a negligible improvement in κ_c.

In contrast, κ_c of hBN35/LLDPE composite increased over a broader range of removed skin layer thickness, i.e., 0–1.51 mm. This was ascribed to the viscosity difference between the two composite types. For instance, the HDPE composites, being more viscose, were expected to be relatively more resistant to compression molding induced hBN orientations. As a result, the hBN/HDPE composites displayed a smaller skin layer thickness compared to the hBN/LLDPE composites. It is worth noting that the removed skin layer thicknesses of the 39 vol. % hBN/HDPE, after which there was no significant change in κ_c was ~0.7 mm. This skin layer thickness was approximately equal to the thickness of the 0.5 mm thick composites, justifying their perpendicular hBN orientation and deteriorated κ.

The extruded composites are formed in a circular strand shape. The SEM images of these strands displayed hBN orientations parallel to their outer surface and the extrusion flow direction. Thus, pressing these strands resulted in hBN orientation that was detrimental to κ_c. Therefore, to exploit this hBN orientation, these strands were pelletized into thin pellets of ~0.5 mm thickness. And when pressed,

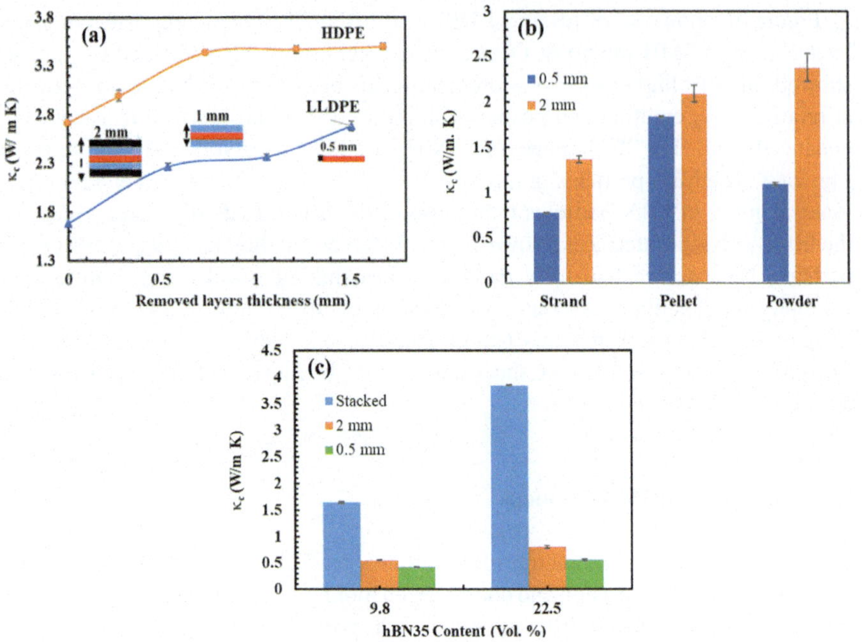

Fig. 5 **a** κ_c of 39 Vol. % hBN35/LLDPE and hBN35/HDPE composites at different removed skin thickness and **b** κ_c of 30 Vol. % hBN35/HDPE composites prepared by pelletizing and powdering circular extrudates

they lay perpendicular to the direction in which the original strands were initially pressed. In addition, to achieve random hBN orientation, the strands were powdered using a simple kitchen grinder. Figure 5b shows κ_c of 0.5-and 2-mm thick 30 vol. % hBN35/HDPE composites prepared from the original strands, pelletized strands, and powdered strands. In general, the pelletized and powdered composites exhibited enhanced κ_c compared to their original strand counterparts. In particular, the 0.5- and 2-mm thick pellets demonstrated 137 and 53% improvement in κ_c over their strand peers, respectively.

Similarly, the 0.5- and 2-mm thick powders demonstrated 41 and 74% improvement in κ_c over their strand equivalents, respectively. The powdered composites exhibited smaller skin layer thickness of only ~100 μm, compared to ~0.7 mm of the strand composites, justifying their enriched κ_c. And the pellets exploited the extrusion-induced hBN orientation resulting in higher κ_c. However, how the 2 mm powdered composite outperformed the 2 mm pelletized composite was not completely understood. Perhaps some pellets, with a cross-sectional diameter of ~3 mm slightly bigger than the compression molding spacer thickness of 2 mm, were pressed in a direction similar to their original strands, weakening κ_c. This interpretation was supported by the higher κ_c of the 0.5 mm pelletized composite compared to the 0.5 mm powdered composite. During pressing of the 0.5 mm thick sheets, it was made sure that most pellets rested on their cross-section.

To exploit the effect of hBN alignment, hBN was aligned following and slightly modifying the multilayer sheet stacking and rotation method proposed by Yang's group [50]. The slight modification was done by stacking only hBN/LLDPE composite sheets of 0.3–0.5 mm thickness instead of alternating hBN/HDPE and LDPE sheets used in their work. The choice of LLDPE composites over HDPE was for their lower viscosity and ease of pressing, especially at high hBN content. Figure 5c shows the thermal conductivities of these composites (Stacked) and of 0.5- and 2-mm thick hBN/LLDPE composites prepared by the conventional method. Here, the 0.5- and 2-mm composites were considered to have perpendicular and random hBN orientations. As anticipated, the stacked composites, having parallel hBN orientation, exhibited remarkably higher κ_c than the 0.5- and 2-mm composites. For instance, at 22.5 vol. % h-BN loadings, the stacked composite showed approximately ~600 and 400% improvement over κ_c of the 0.5- and 2-mm thick hBN/LLDPE, respectively. In addition, κ_c of the stacked composite at 9.8 vol. % hBN loading was 3 and 2 times that of the 0.5- and 2-mm thick sheets at 22.5 vol. % loading, respectively. This shows how critical role hBN orientation plays in the resultant κ_c.

3.2.4 Effect of hBN Exfoliation

To study the effect of filler treatment, hBN35 was BM for 45 and 90 min in the presence of LLDPE at a ratio of hBN to LLDPE of 4:1. Afterward, the milled hBN and LLDPE mixture (M-hBN) was used to prepare M-hBN/LLDPE nanocomposites. Figure 6 shows the thermal conductivities of these composites, in which the hBN35 was milled for 0 (bulk hBN35), 45 min, and 90 min. Attributed to the exfoliation treatment, composites filled with the M-hBN35 displayed improved κ_c than those filled with the bulk hBN35. Moreover, κ_c showed functionality with milling time, which can be related to the extent of the exfoliation treatment. Also, the prior presence of LLDPE in the M-hBN35 improved the processing of their composites, leading to better dispersion, which played an essential role in enhancing κ_c. For example, at 30 vol. %, κ_c of 0-, 45-, and 90-min M-hBN35/LLDPE nanocomposites were 1.03, 1.28, and 1.46 W m^{-1} K^{-1}, respectively, which was about 282, 376, and 442% higher than that of pure LLDPE, respectively.

3.3 *Thermal Stability Analysis*

The non-oxidative thermal stability of hBN/PE composites was thoroughly investigated via TGA. Figure 7a shows the TGA thermograms of hBN35/LLDPE composites with different hBN loadings. The thermal stability of the composites was enhanced by the incorporation of hBN as indicated by T5 and T50 (the temperature corresponding to 5 and 50 wt.% loss) increase with hBN loading, Table 1. For instance, the T5 and T50 of pure LLDPE were 449 °C and 479 °C, and that of 42 wt.

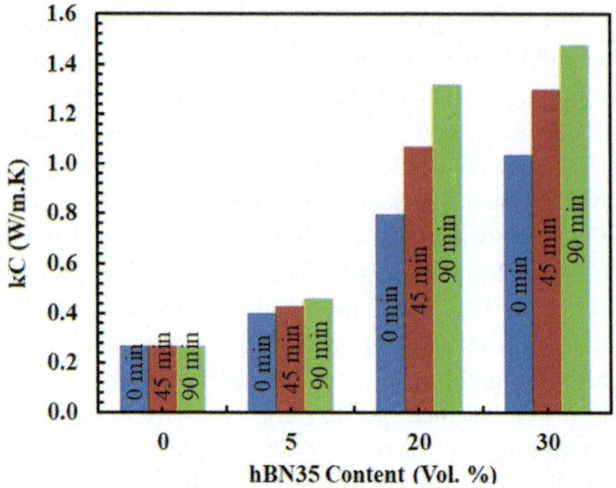

Fig. 6 κ_C of hBN35/LLDPE composites, in which the hBN35 was milled for 0 (bulk hBN35), 45 min, and 90 min

% hBN35/LLDPE were 452 °C and 489 °C, respectively. This was expected knowing that hBN possesses remarkable thermal stability and unusual oxidation resistance, as shown in Fig. 7a by its stability in the studied temperature range. Ayoab et al. [51] reported that the thermal decomposition of hBN/PE composites was enhanced upon the incorporation of hBN. Similarly, Tanimoto et al. [52] stated that the thermal stability of polyimides improved with hBN addition.

Figure 7b compares the thermal decomposition behaviors of hBN35/LLDPE composites prepared by MB and BM. The composites prepared by the MB method showed higher thermal decomposition temperature at all hBN loading than those processed by BM. For instance, the T5 temperature of the MB composites was 446 °C (20.4 wt. %) and 452 °C (42 wt. %), whereas that of the BM composites was 426 °C (18.6 wt. %) and 436 °C (37.7 wt. %). The inferior thermal stability of the BM composites was attributed to the slight exfoliation and oxidation of the hBN due to the BM process, as discussed later.

Figure 7c compares the thermal decomposition of hBN6/LLDPE and hBN35/LLDPE composites. In general, the decomposition temperature of hBN/LLDPE composites increased with hBN loading for both hBN particle sizes. However, among the two-particle sizes, hBN6 showed slightly better enhancement in thermal stability. For instance, the T5 of 52 wt. % hBN6/LLDPE was 456 °C, whereas that of 54 wt. % hBN35/LLDPE was 452 °C. Their better dispersion might cause the higher performance of hBN6. Jung et al. [53] reported higher decomposition temperature of nano-hBN/HDPE relative to micro-hBN/HDPE composites. Similarly, Alaei et al. [54] indicated that the thermal stability of silica/maleic anhydride grafted polypropylene composites had deteriorated when the silica particle size was increased.

Figure 7d shows the TGA curves of pure LLDPE, 42 wt. % hBN35/LLDPE, and 44 wt. % M-hBN35/LLDPE. The M-hBN35/LLDPE manifested a lower onset temperature. For example, the M-hBN35/LLDPE composites' onset temperature was ~80 °C, whereas that of the pure LLDPE was ~369 °C, while, at 369 °C, this composite had lost 1% of their weight. The T5 temperatures of pure LLDPE, 42

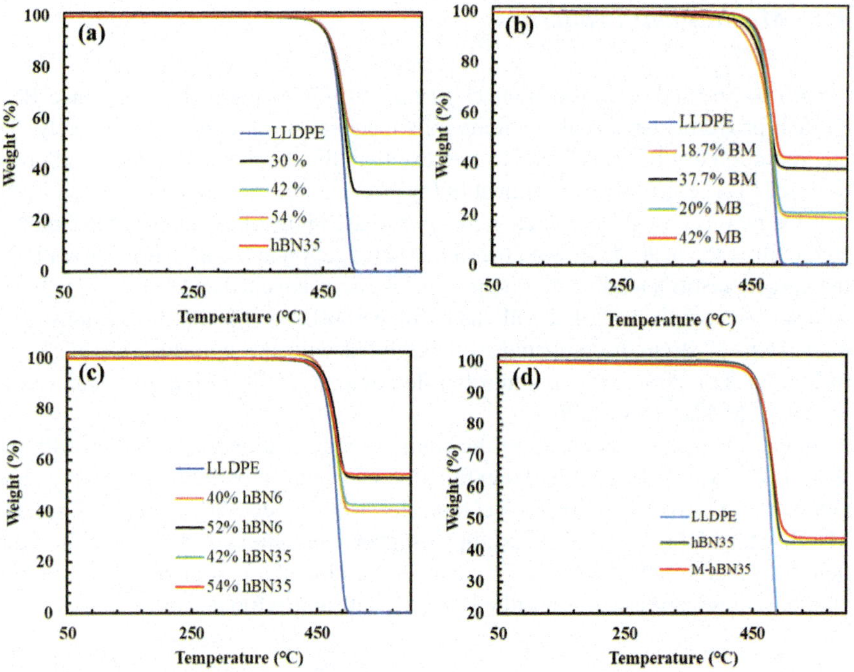

Fig. 7 TGA curves of **a** hBN35 and hBN35/LLDPE composites prepared by MB method, **b** hBN35/LLDPE composites prepared by BM and MB methods, **c** hBN6 and hBN35/LLDPE composites prepared by MB method, and **d** ~22.5 vol. % hBN35 and milled hBN35/LLDPE composited prepared by MB method

Table 1 T_5 and T_{50} of pure LLDPE and LLDPE composites

	LLDPE	42% hBN35	43.9% M-hBN35
T_5 (°C)	449	452	450.3
T_{50} (°C)	479	489	494.8

wt. % hBN35/LLDPE, and 44 wt. % M-hBN35/LLDPE were 449, 452, and 450 °C, respectively. The lower T5 and the higher weight loss of the M-hBN35/LLDPE composites were attributed to the oxidation of the M-hBN35, as confirmed by EDS and XPS. Moreover, the weight loss was ascribed to absorbed moisture evaporation since the onset temperature was around 100 °C. Yu et al. [55] reported a small amount of weight loss of hydroxyl functionalized hBN below 200 °C, which they assigned to the evaporation of moisture and unstable oxygen groups decomposition. Riaz et al. [56] stated that the TGA curve of graphene oxide showed 14.5 wt. % loss between the temperature values of 28 °C and 150 °C, which they accredited to moisture's desorption.

3.4 Mechanical Properties

For PE to be used as a heat transfer surface, it is indispensable to enhance PE's thermal properties and vital to improve its mechanical properties. Fortunately, the incorporation of hBN was found to improve both the thermal and mechanical properties of PE (i.e. LLDPE), as shown in Fig. 8a. The tensile strength increased, and the elongation at break decreased with the increase in the hBN loading. Attributed to its ductility, the neat PE showed the highest elongation at break. However, with the increase in the hBN loading the composite ductility weakened and broke at low strain values. The composites filled with the exfoliated hBN35 (M-hBN35) demonstrated higher tensile strength compared to those filled with the bulk hBN35, as shown in Fig. 8a. For instance, maximum tensile strength of 18.7 MPa was achieved in 25 vol. % M-hBN35/LLDPE.

Figure 8b–d study the effect of the composite fabrication methods (i.e., MB and BM), hBN particle size, and hBN exfoliation on Young's modulus, yield stress, and elongation at break (%) of hBN/PE composites, respectively. At all hBN loadings, the M-hBN35/LLDPE composites exhibited higher yield stress and Young's modulus than the other composite types. In contrast, the elongation at break (%) displayed similar values in the composites filled with the bulk and exfoliated hBN35, which was

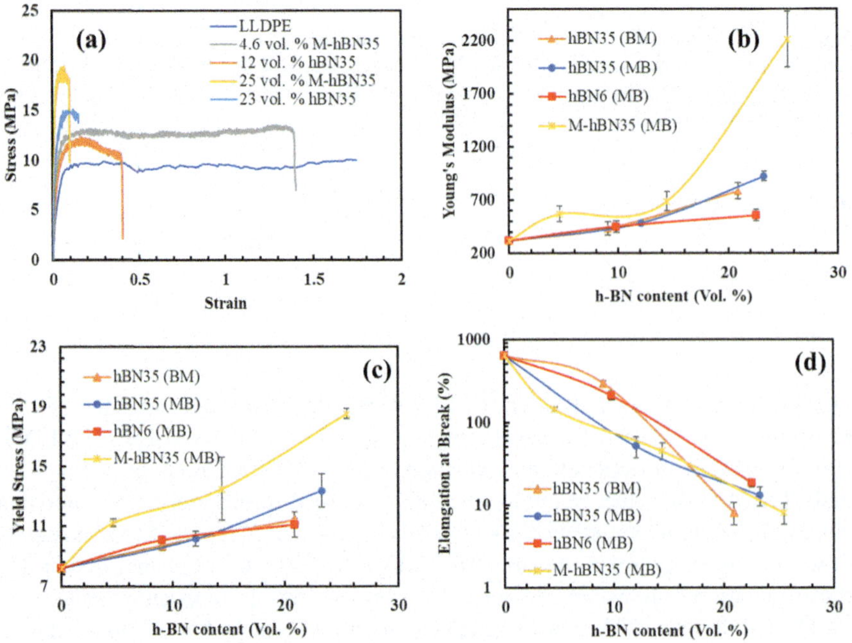

Fig. 8 Stress–strain curves of hBN35/PE and M-hBN35/PE (MB) composites at different hBN loading (**a**), and Young modulus (**b**), Yield stress (**c**), and Elongation at break (%) (**d**) of hBN35/PE (BM), hBN35/PE (MB), M-hBN35/PE (MB), and hBN6/PE (MB)

slightly lower than that of the other composite types. The hBN exfoliation justified the improved tensile properties of M-hBN35/LLDPE composites.

The tensile properties showed slight dependence on the processing method. At 9.8 vol. % both processing methods resulted in similar Young's modulus, yield stress, and elongation at break (%). However, at ~22.5 vol. %, the MB method presented minorly higher yield stress and lower elongation at break (%). Those improvements might be attributed to their hBN orientation. For example, it has been shown earlier that these composites exhibited a skin layer with hBN alignment parallel to the tensile stress direction. Kim et al. [45] reported lower tensile properties in the composites prepared by a powder mixing method than those prepared by a melt blending method.

Similarly, the tensile properties showed negligible functionality with particle size. At low hBN loading (~9.8 vol. %), no clear difference was observed in Young's modulus and yield stress. However, at high hBN loading (~22.5 vol. %), the composites filled with hBN35 displayed slightly better Young's Modulus and yield stress. On the other hand, the elongation at break of the composites filled with hBN6 was always higher than those filled with hBN35. As with the thermal conductivity, the formation of filler networks appeared to enhance the tensile strength of the composites filled with hBN35. Cheewawuttipong et al. [57] reported higher mechanical properties of hBN/ polypropylene composites filled with a big hBN particle size than those filled with a small hBN particle size due to the formation of filler networks. Conclusion.

In this work, the effect of hBN content, particle size, thickness, and alignment, polymer matrix type, and composites preparation techniques on κ_c was meticulously studied. Due to the plate-like geometry, hBN orientation was observed to function as a composite preparation procedure, such as the composite blending method and the composite sheet thickness. This led to composites with hBN mainly aligned orthogonal to the laser flash measurement direction, resulting in deteriorated κ_c. For instance, composites prepared by the MB method displayed skin and core layer structures with orthogonal and parallel hBN orientations, respectively. The thickness of the core layer depended on the composite sheet thickness, which completely disappeared at around 0.5 mm sheet thickness. Therefore, the 0.5 mm thick composites displayed significantly low κ_c. For example, at 39.5 vol. % hBN content, the κ_c values of the 0.5- and 2-mm thick hBN/LLDPE sheets prepared by MB were 1.07 and 1.8 W m^{-1} K^{-1}, respectively. The BM composites showed slightly enhanced κ_c because of having randomly arranged hBN. To prove this, the extrudates were pelletized and powdered. And these resulted in a noticeable improvement in $\kappa_{c,}$ reaching up to 137 and 41% over their original composites peers, respectively. To utilize the effect of alignment, LLDPE composites with perfectly aligned hBN were synthesized. These composites exhibited significantly enriched κ_c, at 22.5 vol. % hBN content reached up to ~600 and 400% improvement over that of the 0.5- and 2-mm thick sheets, respectively. The exfoliation of hBN through the BM method induced some defects and processing difficulty, which was mitigated by adding LLDPE powder during the hBN treatment. This mitigation step was found to improve the composite processing leading to an enhanced κ_c. In addition, the exfoliation was proven highly beneficial in enriching the tensile properties of the PE composites. Owing to the superb thermal and antioxidative stability of hBN, its incorporation thermally stabilized the PE composites making them more suitable for thermal management applications.

Acknowledgements This publication was made possible by an NPRP award [NPRP # 10-0205-170349] from the Qatar National Research Fund (a member of Qatar Foundation). The statements made herein are solely the responsibility of the authors. Special thanks to Qatar Environmental and Energy Research Institute (QEERI) for imaging and characterization.

References

1. Wang X, Wu P (2017) Preparation of highly thermally conductive polymer composite at low filler content via a self-assembly process between polystyrene microspheres and boron nitride nanosheets. ACS Appl Mater Interfaces 9(23):19934–19944. https://doi.org/10.1021/acsami.7b04768
2. Chen H et al (2016) Thermal conductivity of polymer-based composites: fundamentals and applications. Prog Polym Sci 59:41–85. https://doi.org/10.1016/j.progpolymsci.2016.03.001
3. Bigg DM (1995) Thermal conductivity of heterophase polymer compositions. Adv Polym Sci 119:1–30. https://doi.org/10.1007/BFb0021279
4. Burger N, Laachachi A, Ferriol M, Lutz M, Toniazzo V, Ruch D (2016) Review of thermal conductivity in composites: mechanisms, parameters and theory. Progress Polymer Sci 61:1–28, https://doi.org/10.1016/j.progpolymsci.2016.05.001
5. Hussain ARJ, Alahyari AA, Eastman SA, Thibaud-Erkey C, Johnston S, Sobkowicz MJ (2017) Review of polymers for heat exchanger applications: factors concerning thermal conductivity. Appl Therm Eng 113:1118–1127. https://doi.org/10.1016/j.applthermaleng.2016.11.041
6. Huang X, Jiang P, Tanaka T (2011) A review of dielectric polymer composites with high thermal conductivity. IEEE Electr Insul Mag 27(4):8–16. https://doi.org/10.1109/MEI.2011.5954064
7. Xu X, Chen J, Zhou J, Li B (2018) Thermal conductivity of polymers and their nanocomposites. Adv Mater 30(17):1705544–1705554. https://doi.org/10.1002/adma.201705544
8. Mehra N et al (2018) Thermal transport in polymeric materials and across composite interfaces. Appl Mater Today 12:92–130. https://doi.org/10.1016/j.apmt.2018.04.004
9. Huang C, Qian X, Yang R (2018) Thermal conductivity of polymers and polymer nanocomposites. Mater Sci Eng R Rep 132(July):1–22. https://doi.org/10.1016/j.mser.2018.06.002
10. Robbins AB, Minnich AJ (2015) Crystalline polymers with exceptionally low thermal conductivity studied using molecular dynamics. Appl Phys Lett 107(20), https://doi.org/10.1063/1.4936195
11. Papanicolaou GC, Theocaris PS (1979) Thermal properties and volume fraction of the boundary interphase in metal-filled epoxies. Colloid Polymer Sci 257(3):239–246, https://doi.org/10.1007/BF01382364
12. Ye C-M, Shentu B-Q, Weng Z-X (2005) Thermal conductivity of high density polyethylene filled with graphite. J Appl Polym Sci 101(6):3806–3810. https://doi.org/10.1002/app.24044
13. Noh YJ, Kim SY (2015) Synergistic improvement of thermal conductivity in polymer composites filled with pitch based carbon fiber and graphene nanoplatelets. Polym Testing 45:132–138. https://doi.org/10.1016/j.polymertesting.2015.06.003
14. Jiao J, Cui Y, Xia Y (2017) Improved thermal conductivity of epoxy composites prepared with a mixed filler of multiwalled carbon nanotubes and aluminum nitride particles. High Perform Polym 29(4):484–492. https://doi.org/10.1177/0954008316644036
15. Kargar F et al (2018) Thermal percolation threshold and thermal properties of composites with graphene and boron nitride fillers. ACS Appl Mater Interfaces 10(43):37555–37565
16. Xu Y, Chung DDL, Mroz C (2001) Thermally conducting aluminum nitride polymer-matrix composites. Compos Part A: Appl Sci Manuf 32(12):1749–1757, https://doi.org/10.1016/S1359-835X(01)00023-9

17. Zhou W, Wang C, Ai T, Wu K, Zhao F, Gu H (2009) A novel fiber-reinforced polyethylene composite with added silicon nitride particles for enhanced thermal conductivity. Compos Part A: Appl Sci Manuf 40(6): 830–836, https://doi.org/10.1016/j.compositesa.2009.04.005
18. Song J, Zhang Y (2020) Vertically aligned silicon carbide nanowires/reduced graphene oxide networks for enhancing the thermal conductivity of silicone rubber composites. Compos Part A: Appl Sci Manuf 133:105873, https://doi.org/10.1016/j.compositesa.2020.105873
19. Shin YK, Lee WS, Yoo MJ, Kim ES (2013) Effect of BN filler on thermal properties of HDPE matrix composites. Ceram Int 39:S569–S573. https://doi.org/10.1016/j.ceramint.2012.10.137
20. Lee GW, Park M, Kim J, Lee JI, Yoon HG (2006) Enhanced thermal conductivity of polymer composites filled with hybrid filler. Compos A Appl Sci Manuf 37(5):727–734. https://doi.org/10.1016/j.compositesa.2005.07.006
21. Fang H, Bai S-L, Wong CP (2016) "White graphene"– hexagonal boron nitride based polymeric composites and their application in thermal management. Compos Commun 2:19–24. https://doi.org/10.1016/j.coco.2016.10.002
22. Zhang K, Feng Y, Wang F, Yang Z, Wang J (2017) Two dimensional hexagonal boron nitride (2D-hBN): Synthesis, properties and applications. J Mater Chem C 5(46):11992–12022. https://doi.org/10.1039/c7tc04300g
23. Guerra V, Wan C, McNally T (2019) Progress in materials science thermal conductivity of 2D nano-structured boron nitride (BN) and its composites with polymers. Prog Mater Sci 100:170–186. https://doi.org/10.1016/j.pmatsci.2018.10.002
24. Li LH, Chen Y (2016) Atomically thin boron nitride: unique properties and applications. Adv Func Mater 26(16):2594–2608. https://doi.org/10.1002/adfm.201504606
25. Lee J et al. (2016) Boron Nitride Nanosheets (BNNSs) Chemically Modified by "Grafting-From" Polymerization of Poly(caprolactone) for thermally conductive polymer composites. Chem Asian J 11(13):1921–1928, https://doi.org/10.1002/asia.201600470
26. Guo Y, Leung SN (2018) Thermally conductive polybutylene terephthalate/hexagonal boron nitride composites with bimodal fi ller size distribution. Mater Chem Phys 214:221–228. https://doi.org/10.1016/j.matchemphys.2018.04.073
27. Bao J, Edwards M, Huang S, Zhang Y, Fu Y (2016) Two-dimensional hexagonal boron nitride as lateral heat spreader in electrically insulating packaging. Phys D: Appl Phys 49:265501–265509, https://doi.org/10.1088/0022-3727/49/26/265501
28. Wang TB, Jin CC, Yang J, Hu CF, Qiu T (2015) Physical and mechanical properties of hexagonal boron nitride ceramic fabricated by pressureless sintering without additive 114(5,):273–277, https://doi.org/10.1179/1743676114Y.0000000226
29. Duan X et al (2016) Review on the properties of hexagonal boron nitride matrix composite ceramics. J Eur Ceram Soc 36(15):3725–3737. https://doi.org/10.1016/j.jeurceramsoc.2016.05.007
30. Lin Z, McNamara A, Liu Y, Moon KS, Wong CP (2014) Exfoliated hexagonal boron nitride-based polymer nanocomposite with enhanced thermal conductivity for electronic encapsulation. Compos Sci Technol 90:123–128, https://doi.org/10.1016/j.compscitech.2013.10.018
31. Rudolph S (1993) Composition and application of coatings based on boron nitride. Interceram 42(5):4–6
32. Charoo MS, Wani MF (2016) Tribological properties of h-BN nanoparticles as lubricant additive on cylinder liner and piston ring. Lubr Sci 29(4):241–254. https://doi.org/10.1002/ls
33. Su C-Y, Wang J-C, Chen C-Y, Chu K, Lin C-K (2019) Spherical composite powder by coupling polymethyl methacrylate and boron nitride via spray drying for cosmetic application. Materials 12:706–716. https://doi.org/10.3390/ma12050706
34. Zhou S, Yang X, Xu X, Dou SX, Du Y, Zhao J (2020) Boron nitride nanotubes for ammonia synthesis: activation by filling transition metals. J Amer Chem Soc 142(1):308–317, https://doi.org/10.1021/jacs.9b10588
35. Liu C et al. (2019) Carbon and boron nitride nanotubes: structure, property and fabrication. ES Mater Manuf 3:2–15, https://doi.org/10.30919/esmm5f199

36. Pawlak Z, Kaldonski T, Pai R, Bayraktar E, Oloyede A (2009) A comparative study on the tribological behaviour of hexagonal boron nitride (h-BN) as lubricating micro-particles—an additive in porous sliding bearings for a car clutch. Wear 267(5–8):1198–1202. https://doi.org/10.1016/j.wear.2008.11.020
37. Yuan C, Duan B, Li L, Xie B, Huang M, Luo X (2015) Thermal conductivity of polymer-based composites with magnetic aligned hexagonal boron nitride platelets. ACS Appl Mater Interfaces 7(23):13000–13006. https://doi.org/10.1021/acsami.5b03007
38. Jiang Y, Shi X, Feng Y, Li S, Zhou X, Xie X (2018) Enhanced thermal conductivity and ideal dielectric properties of epoxy composites containing polymer modified hexagonal boron nitride. Compos A Appl Sci Manuf 107:657–664. https://doi.org/10.1016/j.compositesa.2018.02.016
39. Sun N et al (2018) Hot-pressing induced orientation of boron nitride in polycarbonate composites with enhanced thermal conductivity. Compos A Appl Sci Manuf 110:45–52. https://doi.org/10.1016/j.compositesa.2018.04.010
40. Donnay M, Tzavalas S, Logakis E (2015) Boron nitride filled epoxy with improved thermal conductivity and dielectric breakdown strength. Compos Sci Technol 110:152–158. https://doi.org/10.1016/j.compscitech.2015.02.006
41. Wattanakul K, Manuspiya H, Yanumet N (2011) Thermal conductivity and mechanical properties of BN-filled epoxy composite: Effects of filler content, mixing conditions, and BN agglomerate size. J Compos Mater 45(19):1967–1980. https://doi.org/10.1177/0021998310393297
42. Sato K, Horibe H, Shirai T, Hotta Y, Nakano H, Nagai H (2010) Thermally conductive composite films of hexagonal boron nitride and polyimide with affinity-enhanced interfaces. J Mater Chem 20(14):2749–2752. https://doi.org/10.1039/b924997d
43. Li A, Zhang C, Zhang YF (2017) Thermal conductivity of graphene-polymer composites: Mechanisms, properties, and applications. Polymers 9(9):1–18. https://doi.org/10.3390/polym9090437
44. Xu C, Miao M, Jiang X, Wang X (2018) Thermal conductive composites reinforced via advanced boron nitride nanomaterials. Compos Commun 10:103–109. https://doi.org/10.1016/j.coco.2018.08.002
45. Kim K, Kim J (2016) Core-shell structured BN/PPS composite fi lm for high thermal conductivity with low filler concentration. Compos Sci Technol 134:209–216. https://doi.org/10.1016/j.compscitech.2016.08.024
46. Ghose S et al (2006) Thermal conductivity of ultem TM/carbon nanofiller blends 2. High Perform Polym 18:961–977
47. Vincent M, Giroud T, Clarke A, Eberhardt C (2005) Description and modeling of fiber orientation in injection molding of fiber reinforced thermoplastics. Polymer 46(17):6719–6725. https://doi.org/10.1016/j.polymer.2005.05.026
48. SadAbadi H, Ghasemi M (2007) Effects of some injection molding process parameters on fiber orientation tensor of short glass fiber polystyrene composites (SGF/PS). J Reinf Plast Compos 26(17):1729–1741. https://doi.org/10.1177/0731684407081352
49. Advani SG, Sozer EM (2013) Process modeling in composites manufacturing. Marcel Dekker Inc., New York, pp 1–451
50. Yang S-Y, Huang Y-F, Lei J, Zhu L, Li Z-M (2018) Composites : Part A Enhanced thermal conductivity of polyethylene/boron nitride multilayer sheets through annealing. Compos A 107:135–143. https://doi.org/10.1016/j.compositesa.2017.12.031
51. Ayoob R (2018) Enhanced dielectric properties of polyethylene/hexagonal boron nitride nanocomposites. J Mater Sci 53(5):3427–3442. https://doi.org/10.1007/s10853-017-1786-y
52. Tanimoto M, Yamagata T, Miyata K, Ando S (2013) Anisotropic thermal di ff usivity of hexagonal boron nitride-filled polyimide films: effects of filler particle size, aggregation, orientation, and polymer chain rigidity. ACS 5:4374–4382. https://doi.org/10.1021/am400615z
53. Jung J et al (2010) Preparations and thermal properties of micro- and nano-BN dispersed HDPE composites. Thermochim Acta 499(1–2):8–14. https://doi.org/10.1016/j.tca.2009.10.013

54. Alaei MH et al. (2013) Effect of particle size on thermomechanical properties of particulate polymer composite, https://doi.org/10.1007/s13726-013-0184-9
55. Yu B et al. (2016) Thermal exfoliation of hexagonal boron nitride for effective enhancements on thermal stability, flame retardancy and smoke suppression of epoxy resin nanocomposites via sol–gel process. J Mater Chem A 4(19):7330–7340, https://doi.org/10.1039/C6TA01565D
56. Riaz S, Park S-J (2019) Thermal and mechanical interfacial behaviors of graphene oxide-reinforced epoxy composites cured by thermal latent catalyst. Materials 12:1354–1357
57. Cheewawuttipong W, Fuoka D, Tanoue S, Uematsu H, Iemoto Y (2013) Thermal and mechanical properties of polypropylene/boron nitride composites. Energy Procedia 34:808–817. https://doi.org/10.1016/j.egypro.2013.06.817

Thermal Conductive Composites Reinforced with Advanced Micro and Nano-sized Boron Nitride Particles

Alok Agrawal and Alok Satapathy

Abstract It is well known that polymers are good thermal insulators and hence they find very restricted applications in areas where good thermal conductivity is highly essential for effective heat dissipation. One such situation is encountered in micro-electronics where substantial heat is generated due to miniaturization and high-degree integration of electronic devices. It becomes desirable for any polymer component in such applications to be thermally conductive and this can possibly be done by reinforcing them with conductive filler particles. Desirable requirements of the polymer for these kinds of usage are low coefficient of thermal expansion (CTE) and high electrical insulation apart from good heat conduction capability. Among the various fillers, conductive ceramics have enormous potential as polymer/ ceramic combinations can fulfil the various requirement of material to be used in microelectronic applications. Boron nitride with its high thermal conductivity and low dielectric constant is considered to be an ideal filler material for polymeric resin. In view of this, the present chapter focusses on the heat conduction behaviour of polymer composites reinforced with micro and nano-sized boron nitride particles as filler. The main highlights of the chapter are the different ways adopted for enhancing the heat conduction capability of this kind of composites.

1 Introduction

For the last two decades, heat dissipation in an electronic device has consistently remained a critical issue and challenging problem. The reason behind this is the miniaturization and the rapidly increasing power density of such devices. Few examples are next-generation integrated circuits, three-dimensional integrated circuits and

A. Agrawal (✉)
Department of Mechanical Engineering, Sagar Institute of Research and Technology, Bhopal 462041, Madhya Pradesh, India
e-mail: alokag03@gmail.com

A. Satapathy
Department of Mechanical Engineering, National Institute of Technology Rourkela, Rourkela 769008, Odisha, India

M. Jawaid and A. Khan (eds.), *Multifunctional Boron-Nitride Composites*, Composites Science and Technology, https://doi.org/10.1007/978-981-99-2866-8_4

high-power density transistors [1]. With an increase in integration of components, heat generation increases. The heat generated needs to be dissipated with high efficiency as is govern the overall performance of the electronic devices. Insufficient heat dissipation capability of electronic devices results in the malfunctioning of such devices and hence reduces their life span. Polymers are extensively used in different microelectronic applications mainly due to their low density, low cost, high electrical insulation and facile processibility [2]. But polymers possess very low thermal conductivity (generally below 0.4 W/m–K) and is inadequate to fulfil heat dissipation requirements of electronic industries. Hence, the development of polymers with high thermal conductivity is the current demand of the market. The addition of thermally conductive fillers in polymers had fascinated consideration as heat dissipative material. The heat conduction capability of such material depends upon many factors like intrinsic thermal conductivity of two phases, loading level of filler, geometry of the filler, arrangement of fillers in polymers, interaction/compatibility between filler and polymer. Even a hybrid combination of fillers of different types or different sizes can be used to obtain synergism in terms of properties [3].

As already mentioned, ceramic fillers are quite suitable for designing polymer composites for microelectronic applications mostly because ceramics possess the finest collective properties like high heat conduction, low dielectric constant and low CTE. Various ceramic fillers that attracted the attention of the scientific community are aluminium nitride, boron nitride, silicon nitride, aluminium oxide, silicon carbide etc. Each of them had certain pros and cons. Among all the ceramic fillers, boron nitride is best suited as a filler material for such applications. Boron nitride is composed of the alternative covalent bond of boron and nitrogen which originates because of the difference in electronegativity between the two elements [4]. Hence, the covalent bond has few characteristics of ionic bonds [5]. Boron nitride is produced synthetically with different crystal structure using different hybridization mode. The different structures of boron nitride are hexagonal boron nitride (hBN), rhombohedral BN (rBN), cubic BN (cBN) and wurtzite BN (wBN). Among the different crystal structures, hBN and cBN exists in thermodynamically stable whereas, rBN and wBN are present in a metastable state. Again, hBN possesses superior thermal properties as compared to cBN due its graphite like structure. Hence, hBN are of interest among the scientific community. With a unique crystal structure, boron nitride has exceptional properties like high thermal stability as it cannot be oxidized up to 850 °C at ambient condition. Its resistance to heat decomposition temperature in an inert atmosphere is around 2000 °C. It possesses high thermal conductivity of 33 W/m–K along the out-plane direction and 125 W/m–K along the in-plane direction. It has low CTE of 2×10^{-6} m/°C. The electrical resistivity of boron nitride is 1016–1018 Ω-cm with a dielectric constant that varies between 3 to 4. Boron nitride is a chemically stable material that is insoluble in water and non-reactive at room temperature. Boron nitride has excellent lubricating characteristics with a low coefficient of friction of 0.2. Because of a combination of many useful properties, boron nitride has many potential applications [6]. Boron nitride is in use as adsorption materials and catalytic materials. BN is used by various industries for storing hydrogen materials. BN is excellent for dielectric substrate. Due to its antioxidant behaviour, it is also used anticorrosive material. It does not catch fire easily and are in use as flame-retardant materials. It has capability to absorb high intensity radiation and can used be used to protect against harmful radiation. Recent use of boron nitride is in medical and developing polymer composite materials [6]. Among different applications, usage of boron nitride as fillers in polymeric resin gain great interest as the combination provide excellence comprehensive properties. A significantly large volume of work has been performed in the past by various researchers across the world on boron nitride filled polymer composites to achieve high thermal conductivity. The present chapter discusses different methods adopted to enhance the thermal conductivity of polymeric resin by reinforcing boron nitride as filler in it. This helps the scientific community to get a benchmark and fundamental basis to perform further work in a similar area so that boron nitride filled polymeric composites can grow their potential application in industries and more specifically for microelectronic applications.

2 Effect of Boron Nitride Content on Thermal Conductivity of Polymer Composite

Boron nitride has high thermal conductivity and because of the difference in thermal conductivity between boron nitride and polymers, its inclusion in polymers obliviously enhance the effective thermal conductivity of the polymer-particulate system. Among various factors that govern the rate of enhancement of thermal conductivity, the major influence is on the loading level of boron nitride as high loading is required to provide pathways for phonon propagation. With the inclusion of boron nitride in polymeric resin, the heat conduction behaviour of the material increases as a function of boron nitride content. The same has been observed by Zhou et al. [7] when they measured the thermal conductivity of boron nitride filled polyethylene composites as shown in Fig. 1. They found that thermal conductivity increases when boron nitride is added to the polyethylene matrix and the increment are a function of boron nitride content. They reported achieving a thermal conductivity of 0.98 W/m–K for BN content of 30% by volume. Zhi et al. [8] used boron nitride in the form of nanotubes and fabricated samples with a different polymeric resin. The polymers used are poly (vinyl butyral), polystyrene, poly (methyl methacrylate) and poly (ethylene vinyl alcohol). They reported that the thermal conductivity of each polymer improves when boron nitride is added. The thermal conductivity of neat polymer and their composites are presented in Table 1. With filler content less than 40 wt.%, the thermal conductivity obtained is appreciable keeping in mind that they have not used any special technique to align the boron nitride nano-tubes (BNNT).

It is further observed that percentage enhancement in thermal conductivity is different for different polymer for the same filler and same content. It is mainly because interfacial thermal transfer between filler and polymer varies from one polymer to another.

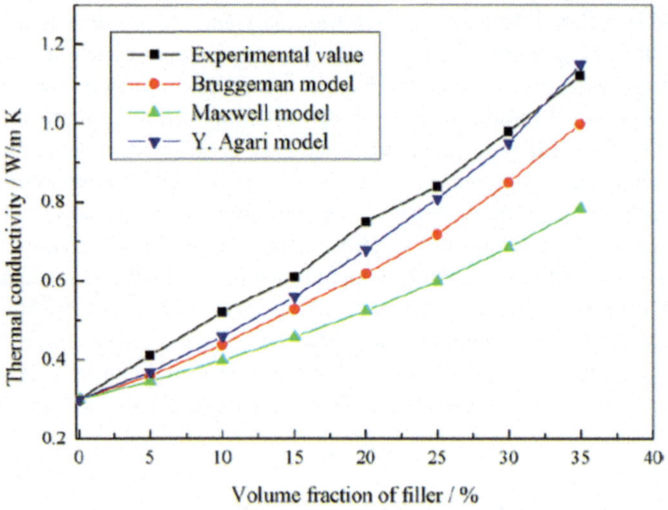

Fig. 1 Thermal conductivity of polyethylene/boron nitride composite as function of filler content [7]

Table 1 Thermal conductivity for different polymer-BNNT composites

Polymer	Thermal conductivity of neat polymer (W/m–K)	Thermal conductivity of polymer/BNNT composite (W/m–K)
poly(vinyl butyral)	0.24	1.81
polystyrene	0.18	3.61
poly(methyl methacrylate)	0.15	3.16
poly(ethylene vinyl alcohol)	0.17	2.50

It is further observed that when the loading level of BN filler is low, the improvement in thermal conductivity of composite is slow. It is mainly because the small number of BN particles randomly dispersed within the matrix having very little interaction among themselves. When the content of BN filler increases beyond a certain value, the interaction between the fillers increases and the formation of a thermally conductive channel established. With this sudden increase in thermal conductivity took place. The phenomenon is known as percolation and for the filler content, this phenomenon took place is called percolation threshold. A similar phenomenon was observed by Gu et al. [9] for a combination of micro-sized boron nitride with polyamide (PI) matrix. In their investigation, for a small mass fraction of 10% BN, a proportionately small increment in thermal conductivity was recorded. The thermal conductivity of pure PI is 0.174 W/m–K increases to 0.254 W/m–K with 10% BN. When the content of BN increases beyond 10%, the rate of increase in thermal

Fig. 2 Thermal conductivity of polytetrafluoroethylene/boron nitride as a function of filler content [10]

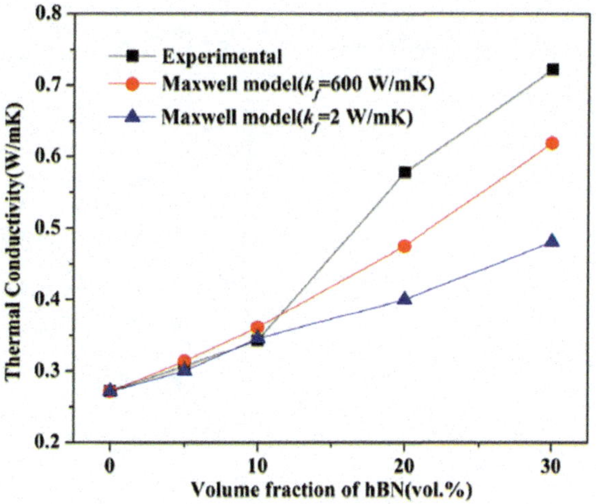

conductivity is high and reaches 0.696 W/m–K with 30% BN content. On a similar note, Pan et al. [10] observed a percolation threshold beyond 10 vol% when they used hexagonal boron nitride polytetrafluoroethylene (PTFE) matrix. The percolation phenomenon observed by them can be seen in Fig. 2.

It is visible that when hBN content increases beyond10 vol%, the slope of the curve increases showing a sudden jump in the value of thermal conductivity beyond that filler content. Again, this percolation threshold is not a constant quantity and can vary with change in either of the phases or with the fabrication route adopted. Hence, it can be concluded from this section that the inclusion of boron nitride will increase the thermal conductivity of the polymeric resin and this increment is a function of boron nitride content. Further, when boron nitride content increases beyond a certain limit, the percolation phenomenon took place results in sudden enhancement in thermal conductivity and helpful to achieve high thermal conductivity polymer.

3 Effect of Change in Ambience Temperature on Thermal Conductivity of Boron Nitride Filled Polymer Composite

It is a well-known fact with change in temperature, the dimension of the material changes. With this change in dimension, different properties of the material tend to change and thermal conductivity is one of them. It has been observed that temperature change has a different effect on thermal conductivity for different material. If we talk about pure polymer, it has been observed that, with an increase in temperature, the thermal conductivity of the polymer increases slightly. It is because the specific heat of the polymer increases with temperature. Against that, the thermal conductivity of polymer composite decreases when temperature increases. Again, this decrease of thermal conductivity is less at low temperature and more when the temperature is

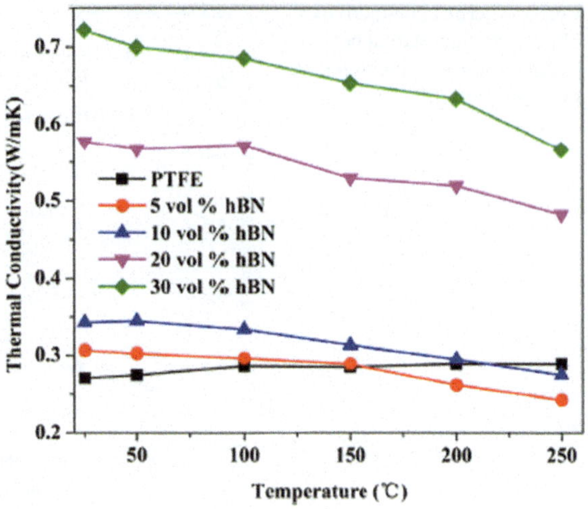

Fig. 3 Thermal conductivity of polytetrafluoroethylene/boron nitride as a function of temperature [10]

high. A similar trend is obtained by Pan et al. [10] when they studied PTFE filled hBN composites and studied the variation in thermal conductivity with temperature for temperature range 25 °C to 250 °C as shown in Fig. 3.

It can be seen from the figure that with an increase in temperature, the thermal conductivity decreases slightly when the temperature is below 100 °C and the decrement is more rapid when the temperature goes beyond 100 °C. The thermal conductivity of composite mainly depends upon the thermal path formed by the conductive fillers. Due to the high coefficient of thermal expansion of the polymer, when temperature increases, composite expands results in a slight increase in the distance between two consecutive conductive particles leading to decreased thermal conductivity [7]. It is seen in this section that with temperature change, the thermal conductivity of the polymer composite filled with boron nitride changes. Thermal conductivity seems to decrease slightly with an increase in temperature. Hence, it can be concluded that as the working temperature of polymer composite increases, the material marginally reduces its tendency to conduct heat.

4 Effect of Boron Nitride Particle Size on Thermal Conductivity of Polymer Composite

It has been seen in the previous section that a major factor that influences the thermal conductivity of polymer composite filled with boron nitride is filler loading. At high filler loading, a thermally conductive path formed and high thermal conductivity is obtained. The size of the filler used is an important factor that governs the maximum possible loading level. The same has been studied by Ishida and Rimdusit [11].

They selected four different size boron nitride particles for studying the packing density. The size of the fillers used by them is 9 μ, 45 μ, 75 μ and 225 μ. In their study, they found that particle with the largest size shown maximum packing density whereas particle with the smallest size shows minimum packing density. The packing density obtained is 60 wt.%, 75 wt.%, 85 wt.% and 88 wt.% for 9 μ, 45 μ, 75 μ and 225 μ particles size respectively. It means that among the four different size fillers, maximum loading of filler is possible when a larger size particle is used and hence the composite with the highest thermal conductivity can be obtained. In their study, they claimed to achieve a very high thermal conductivity of 32.5 W/m–K by reinforcing 88 wt.% (78.5 vol %) boron nitrides in polybenzoxazine matrix as shown in Fig. 4.

Apart from that, at the same loading level, when boron nitride particles of different sizes are embedded in polymer, the thermal conductivity obtained are also found to be different. Zhou et al. [12] reported the effect of different size filler particles on the thermal conductivity of silicon rubber filled with micro size boron nitride particles. In their analysis, they found that for the same filler loading, composite fabricated with bigger particle size filler delivers higher effective thermal conductivity as compared to the composites prepared with smaller particle size filler. They fabricated three sets of composites with three different particle size of 0.3 μ, 6 μ and 20 μ and reported to achieve maximum thermal conductivity for composite fabricated with 20 μ size particles and minimum thermal conductivity with particles size of 0.3 μ.

Kemaloglu et al. [13] studied the effect of BN particle size on thermal conductivity of silicon rubber composites and found that composite prepared with micro-size filler exhibit higher thermal conductivity as compared to composite prepared with nano-size filler. High thermal conductivity is obtained for composite fabricated with bigger particle size is due to formation of higher stability heat conductive path. It is seen that with bigger size particles, the thickness of the conductive path is more and hence had less chance to get disrupted by contacting grains. Further, bigger size particles have lower particle surface area and because of this scattering of phonon is reduced as the interfacial thermal barrier reduced. Also, thermally resistance junctions of polymer are less with bigger size particles as compared to smaller size particles for

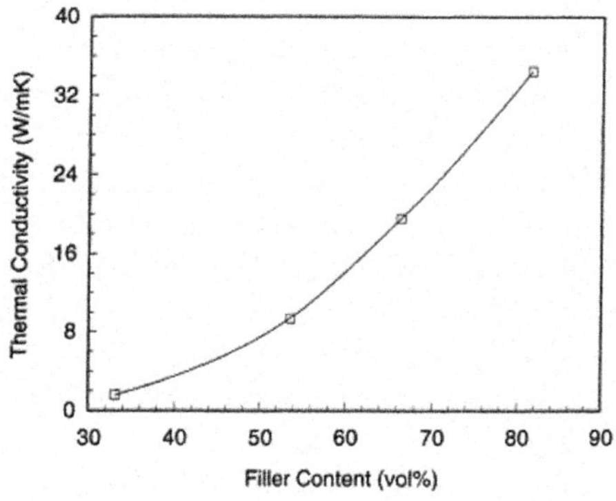

Fig. 4 Thermal conductivity of polybenzoxazine filled boron nitride as a function of filler content [11]

the same filler content and is another reason for achieving high thermal conductivity [12]. Hence, it can be concluded from this section that to achieve high thermal conductivity keeping filler content constant, larger size particles are more suitable.

5 Effect of Surface Modification of Boron Nitride on Thermal Conductivity of Polymer Composite

It has been seen in earlier sections that the thermal conductivity of polymer filled BN increases mainly with an increase in filler content and slightly with an increase in particle size. Though particle size has no appreciable effect on thermal conductivity and increment in thermal conductivity is more dependent on the content of BN. The increment in thermal conductivity is appreciable once the percolation phenomenon took place. But it has to be kept in mind that such high BN loading has some adverse effect as well.

Substantial loading of filler for obtaining high thermal conductivity will result in poor mechanical and processing properties. Also, as the loading of BN in polymer increases, the interfacial area between polymer and BN increases. With an increase in interfacial area, phonon scattering increases which reduce the quantity of heat transfer in the composite material. Phonon scattering will further increase if the interfacial adhesion is not proper and will further reduce the heat conduction behaviour. Hence, improving interfacial adhesion between the BN and polymer provide multiple benefits. Surface treatment of boron nitride filler is known to be an important method for improving the adhesion between filler and polymer and hence reduces the thermal barrier resistance. Surface modification of boron nitride also improves its wettability and helps in a more uniform dispersion of filler in a polymeric resin. With surface treatment, the high surface energy of boron nitride reduces and its amphiphilic nature modifies leading to more uniform dispersion.

In this context, a substantial amount of work has been performed on surface modification of BN filler using different chemicals for imparting more uniform dispersion of particles in the matrix, enhancing the compatibility and reducing the thermal resistance of the interface between filler and matrix. Surface modification of boron nitride using silane is the most widely used method. It is mainly because of its unique chemical structure. Silane has two different chemical structure at its two ends. With two different chemical structure, silane makes the bond with polymer at one end and with boron nitride at another end. Hence, the silane coupling agent acts as a bridge between the polymer and boron nitride and reduces the interfacial thermal contact resistance between the two phases [14]. Among the various silane chemicals, amino silane is the most suitable chemical as it gives higher thermal conductivity as compared to other silane like glycidyl silane. Figure 5 shows the schematic of surface-modified boron nitride with amino silane [15].

Hong et al. [16] studied that, surface modification of boron nitride using silane is useful only when the amount of silane used is optimized. With an increase in silane

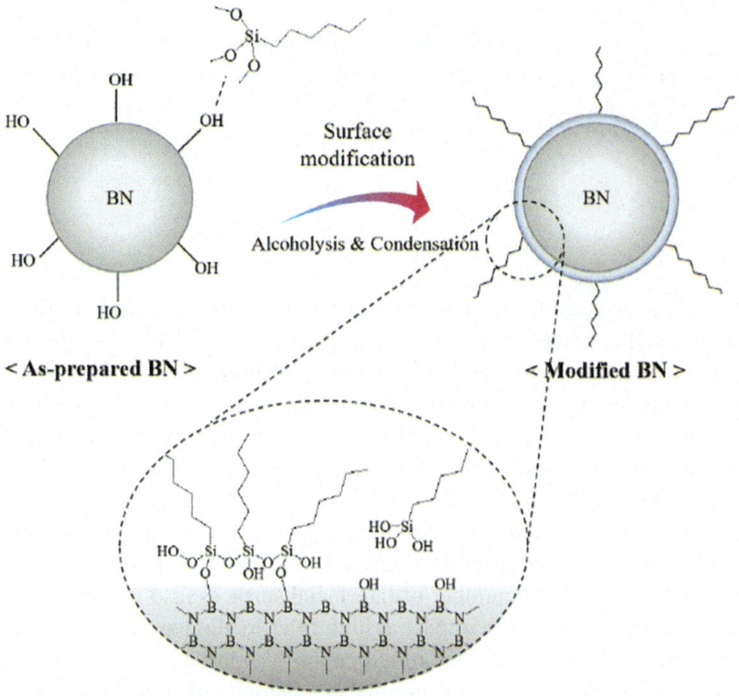

Fig. 5 Schematic of surface modified BN particle with silane coupling agent [15]

content, thermal conductivity initially but an excessive amount reduces the thermal conductivity. They used four different concentration of silane ranging from 2 wt.% to 6 wt.% over boron nitride surface and a fabricated different set of composites with epoxy matrix. Later they measured the thermal conductivity of prepared samples and compared the thermal conductivity value for different silane content for the same filler content of 60 vol %. They found that as the surface of BN is modified with 2 wt.% silane, a slight improvement in thermal conductivity is seen which further increases when 3 wt.% silane is used. With further increase in silane content beyond 3 wt.%, they found that thermal conductivity of composite reduces confirming that excessive silane content reduces the thermal conductivity of the composite. Hence, they concluded that if the thickness of the silane layer was more than a certain value, the thermal conductivity of the composite material decreased, due to the thermal resistance of the silane layer itself.

Thermal conductivity of epoxy/hexagonal boron nitride composite with and without silane treatment was compared by various other researchers as well [17, 18]. All of them reported achieving higher thermal conductivity for composites with silane-modified filler as compared to composite with raw filler. Zhou et al. [17] found that the difference in thermal conductivity of composite between raw BN filler and silane-modified BN filler is less for low filler content and the difference in thermal conductivity increases when the boron nitride content increases. Hou et al. [18] also

Table 2 Percentage of thermal conductivity improvement epoxy composite with raw and modified BN filler

Filler content	Thermal conductivity improvement	
	Raw BN (%)	Silane modified BN (%)
1 wt.%	5.2	13
5 wt.%	41.1	68. 8
10 wt.%	102.1	129.2
30 wt.%	440. 1	513.5

reported a similar trend for the same combination of filler and matrix. They evaluated the thermal conductivity of epoxy composite up to 30 wt.% of boron nitrides and reported that the percentage enhancement in thermal conductivity as a function of untreated and treated filler content. The percentage improvement in thermal conductivity for different filler content achieved by them is shown in Table 2.

From the table, it is visible that at low filler content up to 10 wt.% difference in the improvement of thermal conductivity for composite with treated and untreated filler is less i.e. the maximum difference is around 27%. When filler content increases to 30 wt.%, the maximum difference reaches around 73%. It is because when the filler content increases, the number of interfacial surfaces also increases and hence surface modification has more impact at high filler loading as it reduces interfacial resistance. This shows that surface modification of boron nitride with silane has a huge contribution on improving the thermal conductivity of composite, especially at high filler content.

Huang et al. [19] used a combination of two chemicals for improving the thermal conductivity of boron nitride fillers. They treated the boron nitride first with silane followed by polyhedral oligomeric silsesquioxane (POSS). They measured the thermal conductivity of epoxy composites with raw BNNT and silane-POSS modified BNNT and found that the addition of 30 wt.% of raw BNNT improves the thermal conductivity of composite by around 10 times whereas for the same filler content, thermal conductivity improved by 13.85 times with modified BNNT. Guo et al. [20] also reported earlier to achieve improvement in thermal conductivity using a similar combination of chemicals for surface modification of boron nitride. The chemical used by them is γ-glycidol propyl trimethoxy silane (KH-560) and aminopropyl isobutyl polyhedral oligomeric silsesquioxane (NH2-POSS). They prepared the composite samples with raw BN and surface modified BN with polyamide matrix and measured the thermal conductivity. The effect of surface modification of BN filler on thermal conductivity of composite can be seen in Fig. 6. It is visible from the figure that composite prepared with functionalized boron nitride (f-BN) exhibits higher thermal conductivity as compared to composite prepared with raw BN. The reason for such a difference is the difference in interfacial thermal resistance. When raw BN is used, the interfacial thermal resistance of composite is 0.7357 m^2K/W whereas with surface modified BN interfacial thermal resistance of composite is 0.7057 m^2K/W. Decreased value of thermal resistance is a result of improved thermal conductivity.

Fig. 6 Effect of surface modification of boron nitride on thermal conductivity of composite [20]

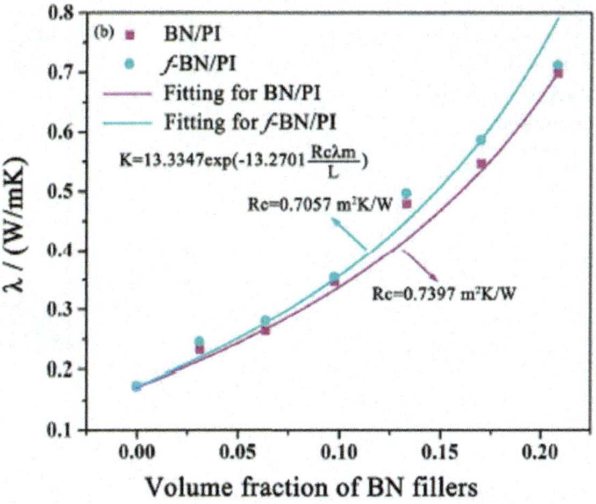

Tereo et al. [21] used catechin for modification of the surface of BNNT for preparing polyvinyl alcohol (PVA) based composite film. Catechin is epicatechin gallate molecules having various aromatic rings. These aromatic rings interact effectively with the BNNT surface via π-π bonding and form a bridge between matrix and filler phase and enhance the heat conduction behaviour between the interface and results in an improvement in effective thermal conductivity. They reported achieving 267% improvement in thermal conductivity for very low catechin modified BNNT content of 3 wt.%. For reducing the interfacial thermal resistance between the hBN and epoxy, Jiang et al. [22] grafted poly(glycidyl methacrylate) (PGMA) chains over the surface of hBN. The grafting of PGMA over hBN results in a strong chemical bonding between the hBN and epoxy. The reported thermal conductivity of epoxy composite with and without surface-modified hBN is shown in Fig. 7. It is visible from the figure that the thermal conductivity of the composite for maximum filler content reaches 1.198 W/mK. However, hBN without surface modification for same content delivers lower thermal conductive composite. The improvement in thermal conductivity with PGMS modified hBN is because of the stronger interfacial interaction due to the longer molecular chain of PGMA.

Pan et al. [23] used tannic acid for the modification of the surface of boron nitride nanosheets. Tannic acid is selected by the group because it contains a substantial amount of catechol and pyrogallol structures. This provides good dispersity of BN nanosheet in the solution. Also, with high surface adhesion characteristics, tannic acid is easily absorbed over the surface of BN nanosheets via π-π bonding. This results in improved interfacial interaction between BN nanosheet and epoxy matrix and hence the improved thermal conductivity is obtained. Kim et al. [24] used benzoic acid for modifying the surface of boron nitride. They reported achieving high thermal conductivity of 2.71 W/m–K for a modified BN content of 50 wt.% which is 1.5 times that received for epoxy composites with raw boron nitride. Modification of BN with benzoic acid results in carboxyl-carboxylate interaction during the

Fig. 7 Effect of surface modification of boron nitride with PGMA grafting on thermal conductivity of composite [22]

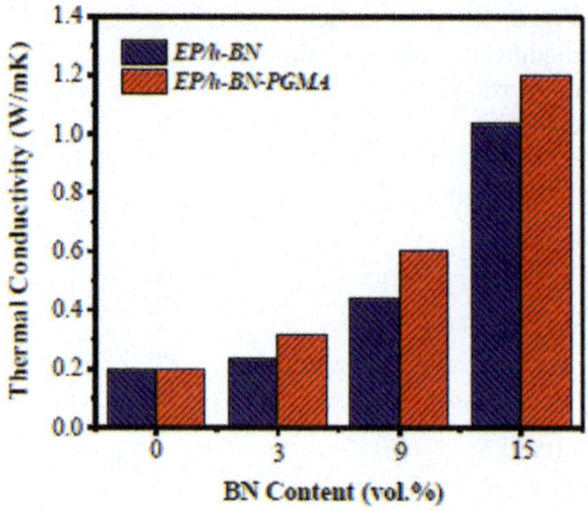

epoxy curing process, resulting in the formation of a continuously thermal conductive pathway, which is responsible for the improved thermal conductivity. Hence, it can be concluded from this section that to achieve high thermal conductivity keeping boron nitride content constant, surface modification of boron nitride play an important role as it improves the interfacial adhesion and reduce interface thermal resistance between polymer and boron nitride.

6 Effect of Hybridization on Thermal Conductivity of Polymer Composite

A hybrid composite is defined as a multi-component composite system, consisting of a matrix phase reinforced with more than one filler not similar in all respect. It means that, if more than one types of fillers are used or one filler type with different particle size or shape are used, all are categorized as hybrid filler. Further, if more than one matrix phase is used in combination with a single filler similar in all respect then it is also a type of hybrid composites. In this section, all such types of hybrid composites are discussed having at least one phase of boron nitride. Li and Hsu [25] fabricated polyamide matrix composite using two different sizes of boron nitride particles. The size of BN particles used by them is 1 μm and 70 nm. Four different sets of composites are fabricated by them. The first set is fabricated with micro-size boron nitride (mBN) particles, the second set is fabricated with nano-sized boron nitride (nBN) particles, third and fourth sets are fabricated with a combination of micro and nano-size filler. In the third set, 70% of particles are of mBN whereas the remaining 30% are of nBN. In the fourth set, 70% particles are of nBN whereas the remaining 30% are mBN. With this combination, composite for three different filler content is prepared and their thermal conductivity is measured. From the measured value of thermal conductivity, they found that a hybrid combination of fillers shows better thermal conductivity. Maximum thermal conductivity is obtained for a combination of 70% mBN and 30% nBN in the polyamide matrix. A similar result of increased thermal conductivity was found by Yi et al. [26] when they used combination of micro and nano-size filler instead of either micro or nano-size filler as shown in Fig. 8. It is visible that around 20% increased value of thermal conductivity is obtained when two different size BN particles are used as compared to single size BN particles. The high thermal conductivity for the combination of micro and nano-size filler is attributed to the formation of the random conductive bridge as the hybrid combination of filler in appropriate ratio when incorporated in matrix system, the thermally conductive network will be formed easily. In this case, the main thermal conductive path in the composite is formed by micro-size BN particles and nano-sized BN particles are used to establish a connection between the micro-particulates. With this mechanism, a more conductive path is formed to facilitate transportation of heat and increases the thermal conductivity of the composite body.

Apart from hybrid combination different particle size of BN, different type of BN when used in hybrid combination can also produce the same effect of improved thermal conductivity in polymeric composites. Fang et al. [27] measured the thermal conductivity of polydimethylsiloxane composite filled with single filler i.e. boron nitride nanosheet and a hybrid filler combination of boron nitride nanosheet and boron nitride foams (BNF) for filler content up to 10 wt.%. The variation in thermal conductivity for two sets of composites are compared and the result obtained by them is shown in Fig. 9. It was found in their investigation that composite prepared with a hybrid combination of fillers shows high thermal conductivity as compared to the

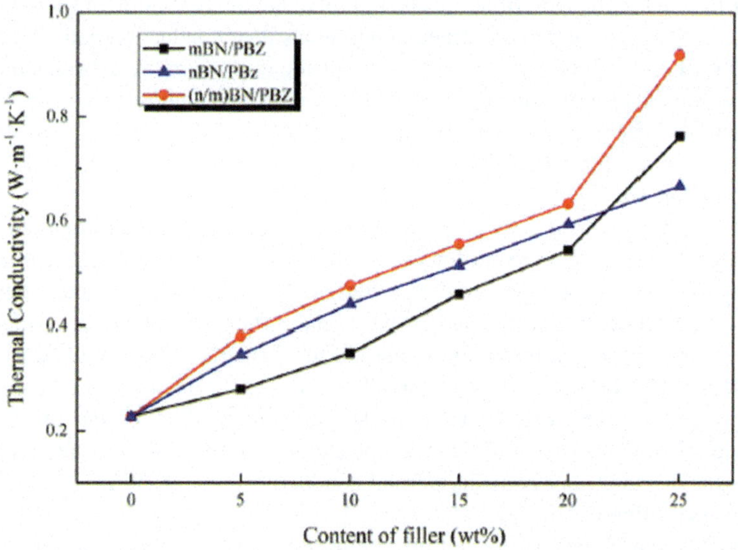

Fig. 8 Effect of different size of BN in hybrid combination on thermal conductivity of polybenzoxazine composite [26]

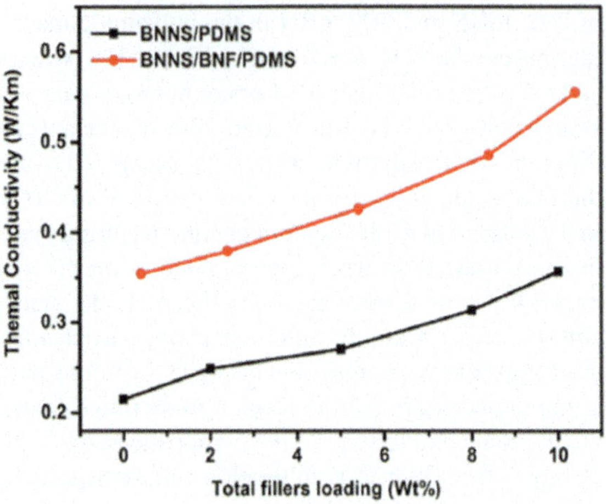

Fig. 9 Effect of different type of BN in hybrid combination on thermal conductivity of polydimethylsiloxane composite [27]

composite filled with only BNNS over a range of filler content. For polydimethylsiloxane filled with BNNS, thermal conductivity is improved by 69.9% for 10 wt.% of filler, whereas, for polydimethylsiloxane filled with 10 wt.% of BNNS/BFS, thermal conductivity is improved by 164.5%. The improvement in thermal conductivity with hybrid filler will be again attributed to the formation of the conduction path in the matrix body.

Liu et al. [28] used a hybrid combination of the matrix with boron nitride for preparing a 3D segregated structure and reported that the formation of such structure is an effective approach to construct a continuous heat conduction path. They used polystyrene and polypropylene with boron nitride to prepare the hybrid composite. The measured thermal conductivity of hybrid composite is compared with the thermal conductivity of composite with single matrix system i.e. PS/BN composite. The difference in thermal conductivity obtained between two composite systems are shown in Fig. 10.

It is visible that tertiary composite possesses greater thermal conductivity as compared to binary composite for every filler content. The difference in thermal conductivity between two sets of composites are less for low filler content but when the filler content increases beyond 20 wt.%, the difference in thermal conductivity increases appreciably. It is mainly because of the formation of thermally conductive networks when the percolation threshold occurs. The network structure becomes stronger as the filler content increases. For maximum filler content of 50 wt.%, the thermal conductivity of tertiary composite is 5.57 W/m–K whereas, for binary composite at same filler content, thermal conductivity is 2.02 W/m–K which is an appreciable difference.

Another way of preparing hybrid composite is using more than one type of filler. Here also the main aim of using such hybrid filler is to establish a thermally conductive pathway for fast conduction of heat. Liang et al. [29] used the combination of boron nitride with aluminium nitride and prepared epoxy composite. They

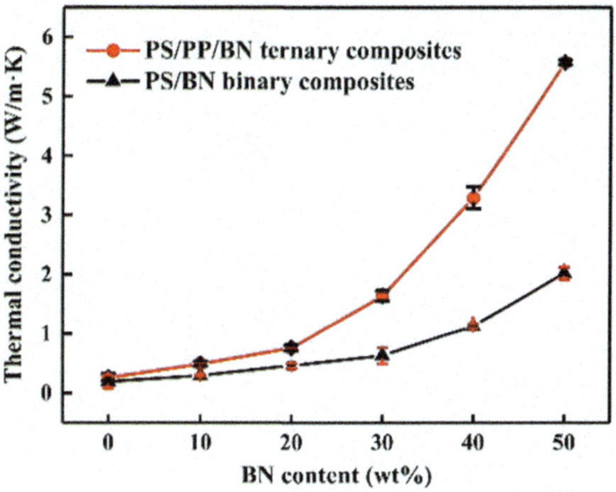

Fig. 10 Difference in thermal conductivity between PS/PP/BN composite and PS/BN composite [28]

reported achieving superior thermal conductivity for hybrid composites as compared to composite prepared with either BN or AlN. For 40 vol.% EP/BN-AlN, thermal conductivity increased to 2.4 W/m–K against 0.3 W/m–K for neat epoxy and 1.3 W/m–K for EP/BN composite. The remarkable increment in thermal conductivity is attributed to the synergetic effect of BN and AlN on the formation of the phonon transmission pathway. They found that mutual interaction between both the particles are excellent and result in uniform dispersion and formation of thermal conductive network with the reduced gap between filler results in very low thermal interface resistance. Guo et al. [30] used a hybrid combination of BN and carbon nanotube (CNT) in polyethylene matrix and reported that the inclusion of CNT along with BN appreciably increases the thermal conductivity of the composite system. In their analysis, they found that thermal conductivity increases with an increase in either of the filler content, though the effect of BN content is more dominating CNT content also plays a very significant role in proper utilization of intrinsic property of BN. When BN content kept constant, and a small amount of CNT is added, the distance between the consecutive BN particles is large enough to cause strong phonon scattering, results in limited improvement in thermal conductivity. When the content of CNT increases, it plays a role of bridge between the BN particles resulting in the formation of a better continuous path and formation of an effective hybrid filler thermal network and because of this thermal conductivity of the system increases. A similar increasing trend in thermal conductivity is obtained for a hybrid combination of boron nitride and graphene when reinforced in the epoxy matrix as reported by Feng et al. [31]. The difference in thermal conductivity obtained with and without graphene is shown in Fig. 11. It is visible from the figure that the inclusion of graphene sheet enhances the thermal conductivity of epoxy/hBN composite appreciably. Further, it can be seen that the difference in thermal conductivity with and without graphene increases with an increase in filler content as observed earlier by Guo et al. [30].

Jin et al. [32] used a combination of graphene nanoplatelets and CNT with boron nitride for the preparation of hybrid composite with polycarbonate matrix and studied

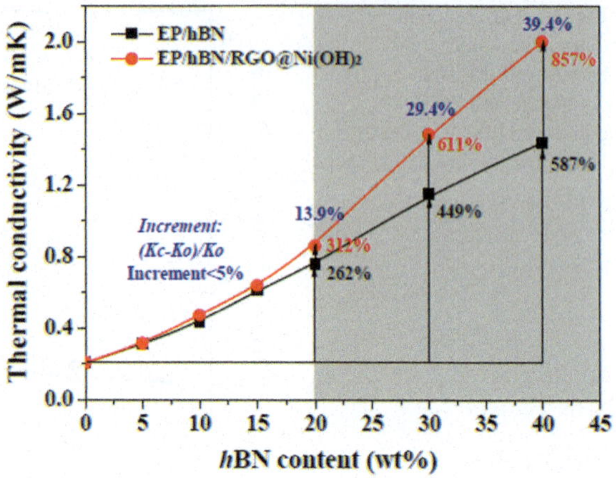

Fig. 11 Effect of hybridization on thermal conductivity of polymer composites [31]

the combined effect of multi-fillers on thermal conductivity of composites. It is seen that the combination of graphene and CNT significantly reduce the thermal contact resistance at the interface between boron nitride and polycarbonate matrix. With reduced thermal interface resistance, the thermal conductivity of the composite increases significantly. The difference in thermal conductivity, with and without graphene/CNT of polycarbonate/BN composite is appreciable mainly due to the difference in information heat conduction path in two different composites. Hence, it can be concluded from this section that to achieve high thermal conductivity, hybrid composites shown more fruitful results as compared to single filler composite mainly because in hybrid composite more structured heat conduction path is formed and reduced phonon scattering is achieved as compared to single filler composite.

7 Effect of Fabrication Method on the Thermal Conductivity of Boron Nitride Filled Polymer Composite

It has been seen in an earlier section that thermal conductivity of polymer composite filled with boron nitride can be improved by increasing BN content, increasing the size of BN particles, reducing interfacial thermal resistance between the two phases by surface modification of BN or by the building of thermally conductive path by hybridization. Among them, it is observed that by forming the conductive path, appreciable enhancement in thermal conductivity of composite can be obtained. Apart from hybridization, a conductive path in polymer composite can also be formed by using different fabrication technique. In this section of the chapter, the effect on thermal conductivity of polymer composite filled with BN by adopting different fabrication techniques will be discussed. It will be seen that by adopting the proper fabricating method, the structured thermal conductive path can be formed and the same will result in high thermally conductive polymer composites.

Zhang et al. [33] used a multistage stretching method for fabricating polyethylene/BN composite and studied the effect of intensity of the shear field on the distribution of BN particles within the matrix body to establish a thermally conductive path. They reported achieving improvement in thermal conductivity of composite when the strong shear field is applied. The effect of the shear field is noticeable once the filler content increases beyond 20 wt.% whereas it has a negligible effect on thermal conductivity for low filler content. At high filler content, without shear field, agglomeration of filler took place and with this, it is difficult to form a conductive path. As the strong shear field is applied, the BN aggregates damaged causes more uniform dispersion of BN particles and results in the formation stable thermal conductive path. Cho et al. [1] worked intending to achieve high thermal conductivity at low BN content. They prepared polysiloxane based composites and controlled the gap between the BN fillers with the help of an applied electric field.

They used DC electric field and AC electric field separately to orient and relocate the BN filler in polysilixane polymer. The thermal conductivity of composites prepared without application of electric field and with the applied electric field was compared and observed that for the entire range of BN content the composite prepared under the electric field exhibited higher thermal conductivity. The highest thermal conductivity is obtained for composite prepared with DC electric field followed by composite prepared with AC electric field. Minimum thermal conductivity is obtained for composite prepared without any electric field. Wang et al. [34] prepared BNNS/polystyrene microsphere using the latex blending method followed by compression moulding to obtained the composite sheets. The BNNS and polystyrene microsphere are mixed and because of strong electrostatic interaction between the two, they form a complex microsphere. Later, the microspheres are compressed at an elevated temperature so that BNNS will be distributed at the interface of polystyrene. Using this fabrication route, homogeneous and oriented distribution of BNNS in polystyrene matrix is obtained and with this unique structure, significant improvement in thermal conductivity is obtained.

Sun et al. [35] used boron nitride plates as filler in polycarbonate matrix and fabricated composites for different filler content via the hot-pressing method. They used the hot-pressing method to align the BN plates in the horizontal direction with polymer (HoBN/PC composite) for providing a low resistance path for phonon transport. They also prepared similar sets of composites with a random distribution of BN plates in polycarbonate matrix (RoBN/PC composites) for comparing the value of thermal conductivity obtained in two different methods. The thermal conductivity of HoBN/PC composites is much higher as compared to RoBN/PC composite for each composition of prepared composites as seen in Fig. 12. The reason for enhancement in thermal conductivity is mainly because of the formation of the more thermal conductive path with oriented BN plates obtained by hot pressing. For maximum BN content of 18.5 vol.%, the thermal conductivity of HoBN/PC composite is 3.09 W/m–K whereas that of RoBN/PC composite is only 0.96 W/m–K. Similarly, the high value of thermal conductivity obtained by Cao et al. [36] when they used the same fabrication technique of hot pressing for aligning the BNNS in polyamide.

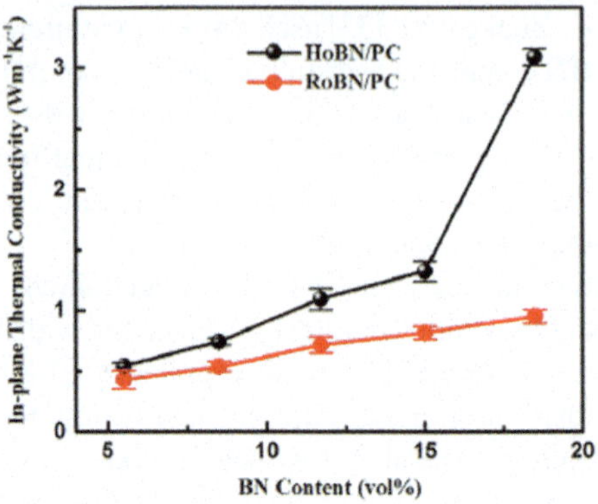

Fig. 12 Thermal conductivity of polycarbonate/BN composites fabricated with and without hot pressing method [35]

Wang et al. [2] established a new technique to form a three-dimensional network of boron nitride in polystyrene for efficiently conducting heat. They used oil–water interface assembly for this purpose. They prepared Pickering emulsion Benzoyl peroxide, styrene oil, deionized water and boron nitride. Benzoyl peroxide and styrene are used to prepare the oil phase and BN is dispersed in deionized water. This water dispersion is mixed with the oil phase and emulsified to get a Pickering solution. Later, with the help of in situ polymerization, they prepared PS@BN microspheres which were then hot compressed to get a composite sheet. Fabricating composite with this technique results in appreciable improvement in thermal conductivity in their work. The improvement in thermal conductivity is attributed to the formation of a conductive 3-D network structure.

Chen et al. [37] was inspired by the aligned network structure of meringue sheets used for the preparation of millefeuille cake and fabricated poly(vinyl alcohol)/BNNS nanocomposite. First, they prepared the poly(vinyl alcohol) (PVA) nanofiber layer covered with BNNS sheet by electrostatic spraying of BNNS sheet over PVA nanofiber. The porous membrane of PVA coated by BNNS is cut into pieces to construct layer by layer mat by lamination. PVA fiber and BNNS layer were placed at the top of the structure and compressed to get the composite body. They also prepared the PVA/BNNS composites by distributing the BNNS randomly over PVA and compared the difference in thermal conductivity noticed for composites fabricated by different methods. From the measured value of thermal conductivity, they found that when BNNS was randomly oriented, the thermal conductivity increases gradually with filler content, whereas in composites with BNNS network structure, a sharp increase in thermal conductivity is noticed. The maximum thermal conductivity obtained is 21.4 W/m–K in their investigation by adopting millefeuille-inspired thermally conductive polymer nanocomposites with overlapping BN nanosheets.

Shang et al. [38] designed a multilayer structure for fabricating composites to achieve high thermal conductivity. First, they melt blended HDPE and hBN platelets and prepared the first thin film using hot-pressing. For preparing the second thin film, they used virgin LDPE. Later, film of HDPE/hBN and LDPE were stacked alternatively and annealed to establish contact between the two layers. The stacked film was cut into small pieces and hot-pressed to obtain a composite body. Another set of composites were prepared by the random distribution of the hBN platelets in polyethene. Thermal conductivity of both sets of composites was measured and the result obtained for filler content up to 15 wt.% are shown in Fig. 13. It can be seen from the figure that the thermal conductivity of multiplayer composites is greater than the thermal conductivity of composites with the random distribution of filler for the same filler loading. Thermal conductivity enhancement of prepared samples as compared to virgin polymer is reported as around 50% and 73% for random PE/BN composites and multilayer PE/BN composites respectively. In multilayer composites, the continuous thermal conductive path is created using a new technique which results in lower phonon scattering and increases the overall thermal conductivity.

Ou et al. [39] used a mechanochemical approach to fabricate polyamide/boron nitride composite film. In this process of fabrication, boron nitride was uniformly dispersed in the solution of poly(amic acid). In this solution, BN platelets and polymer form covalent linkages under the application of mechanical force. With this application of mechanical force, boron nitride platelets orient themselves in a plane inside the polyamide matrix resulting in a well-established heat conduction network. The fabricated composites film is designated as PI-BN-B was undergone thermal conductivity measurement and the result is compared with the thermal conductivity composite prepared with in-situ polymerization (PI-BN-S). The comparison of thermal conductivity obtained in two different cases is shown in Fig. 14.

It is seen from the figure that the thermal conductivity of PI-BN-B composite is higher as compared to their counterpart. The maximum thermal conductivity of PI-BN-B composite is 16.7 W/m–K whereas that of PI-BN-S composite is 11.8 W/m–K. Hence it can be said that fabrication of composite via mechano-chemical results in the formation of the more stable conductive path as compared to composite prepared via in-situ polymerization. Hence, it can be concluded from this section that adaptation of proper fabrication route plays a key factor in the formation of the conductive path by forming different network structure. With the formation of a well-structured thermal conductive network, improvement in thermal conductivity in the composite system is achieved.

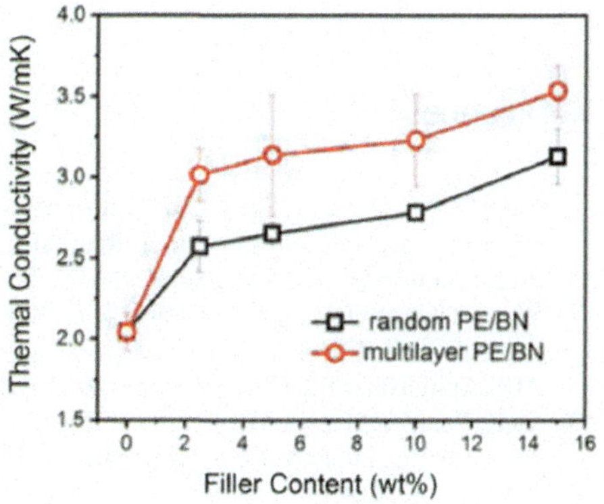

Fig. 13 Comparison of thermal conductivity of random PE/BN composite and multilayer PE/BN composites [38]

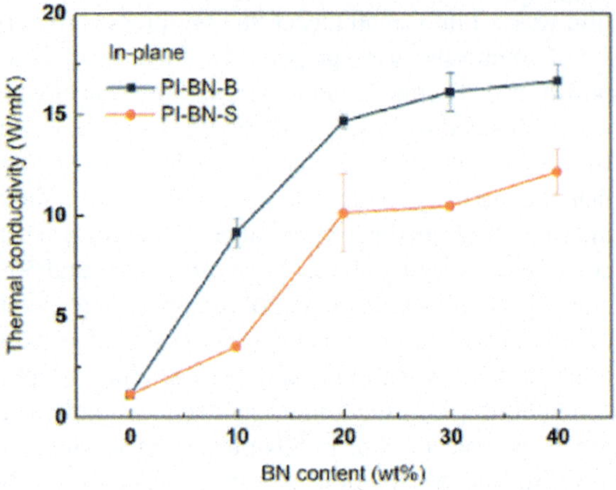

Fig. 14 Effect of fabrication route adopted on thermal conductivity of polyamide/BN composites [39]

8 Conclusions

With the rapidly growing need for complex integration and device miniaturization, electronic components with rapid heat conduction ability are in demand for the efficient discharge of accumulated heat to sustain the device performance. Polymer composites reinforced with appropriate type and number of conductive ceramics have emerged as potential candidates for providing this kind of functionality. In line with this philosophy, the present chapter has been designed to present a comprehensive review focused on the heat conduction behaviour of polymer composites reinforced exclusively with micro and nano-sized boron nitride. It has been reported, based on the available literature, on how conductivity of boron nitride filled polymers depends on various factors like filler content, size and shape. It also shows that the surface modification of boron nitride, hybridization of fillers and favourable fabrication route have remarkable effects on improving the heat conduction behaviour. It can thus be concluded that this class of composites exhibit promising results supporting their candidature for use in micro-electronics in a bigger way in coming days.

References

1. Cho HB, Nakayama T, Suematsu H, Suzuki T, Jiang W, Niihara K, ... Choa YH (2016) Insulating polymer nanocomposites with high-thermal-conduction routes via linear densely packed boron nitride nanosheets. Compos Sci Technol 129:205–213
2. Wang R, Cheng H, Gong Y, Wang F, Ding X, Hu R, ... Tian X (2019) Highly thermally conductive polymer composite originated from assembly of boron nitride at an oil–water interface. ACS Appl Mater Interfaces 11(45):42818–42826
3. Yu K, Yuan T, Zhang S, Bao C (2021) Hypergravity-induced accumulation: a new, efficient, and simple strategy to improve the thermal conductivity of boron nitride filled polymer composites. Polymers 13(3):459

4. Yadav V, Kulshrestha V (2019) Boron nitride: a promising material for proton exchange membranes for energy applications. Nanoscale 11(27):12755–12773
5. Kumar R, Karkamkar A, Bowden M, Autrey T (2019) Solid-state hydrogen rich boron–nitrogen compounds for energy storage. Chem Soc Rev 48(21):5350–5380
6. Pan D, Su F, Liu H, Ma Y, Das R, Hu Q, ... Guo Z (2020) The properties and preparation methods of different boron nitride nanostructures and applications of related nanocomposites. Chem Record 20(11):1314–1337
7. Zhou W, Qi S, An Q, Zhao H, Liu N (2007) Thermal conductivity of boron nitride reinforced polyethylene composites. Mater Res Bull 42(10):1863–1873
8. Zhi C, Bando Y, Terao T, Tang C, Kuwahara H, Golberg D (2009) Towards thermoconductive, electrically insulating polymeric composites with boron nitride nanotubes as fillers. Adv Func Mater 19(12):1857–1862
9. Gu J, Lv Z, Wu Y, Guo Y, Tian L, Qiu H, Zhang Q (2017) Dielectric thermally conductive boron nitride/polyimide composites with outstanding thermal stabilities via in-situ polymerization-electrospinning-hot press method. Compos A Appl Sci Manuf 94:209–216
10. Pan C, Zhang J, Kou K, Zhang Y, Wu G (2018) Investigation of the through-plane thermal conductivity of polymer composites with in-plane oriented hexagonal boron nitride. Int J Heat Mass Transf 120:1–8
11. Ishida H, Rimdusit S (1998) Very high thermal conductivity obtained by boron nitride-filled polybenzoxazine. Thermochim Acta 320(1–2):177–186
12. Zhou T, Qi WY, Zhao SH, HZ, Liu NL (2007) Thermally conductive silicone rubber reinforced with boron nitride particle. Polymer Compos 28(1):23–28
13. Kemaloglu S, Ozkoc G, Aytac A (2010) Properties of thermally conductive micro and nano size boron nitride reinforced silicon rubber composites. Thermochim Acta 499(1–2):40–47
14. Agrawal A, Chandrakar S (2020) Influence of particulate surface treatment on physical, mechanical, thermal, and dielectric behavior of epoxy/hexagonal boron nitride composites. Polym Compos 41(4):1574–1583
15. Jang I, Shin KH, Yang I, Kim H, Kim J, Kim WH, ... Kim JP (2017) Enhancement of thermal conductivity of BN/epoxy composite through surface modification with silane coupling agents. Colloids Surf A: Physicochem Eng Aspects 518:64–72
16. Hong JP, Yoon SW, Hwang TS, Lee YK, Won SH, Nam JD (2010) Interphase control of boron nitride/epoxy composites for high thermal conductivity. Korea-Australia Rheology J 22(4):259–264
17. Zhou W, Zuo J, Zhang X, Zhou A (2014) Thermal, electrical, and mechanical properties of hexagonal boron nitride–reinforced epoxy composites. J Compos Mater 48(20):2517–2526
18. Hou J, Li G, Yang N, Qin L, Grami ME, Zhang Q, ... Qu X (2014) Preparation and characterization of surface modified boron nitride epoxy composites with enhanced thermal conductivity. Rsc Adv 4(83):44282–44290
19. Huang X, Zhi C, Jiang P, Golberg D, Bando Y, Tanaka T (2013) Polyhedral oligosilsesquioxane-modified boron nitride nanotube based epoxy nanocomposites: an ideal dielectric material with high thermal conductivity. Adv Func Mater 23(14):1824–1831
20. Guo Y, Lyu Z, Yang X, Lu Y, Ruan K, Wu Y, ... Gu J (2019) Enhanced thermal conductivities and decreased thermal resistances of functionalized boron nitride/polyimide composites. Compos Part B: Eng 164:732–739
21. Terao T, Bando Y, Mitome M, Zhi C, Tang C, Golberg D (2009) Thermal conductivity improvement of polymer films by catechin-modified boron nitride nanotubes. J Phys Chem C 113(31):13605–13609
22. Jiang Y, Shi X, Feng Y, Li S, Zhou X, Xie X (2018) Enhanced thermal conductivity and ideal dielectric properties of epoxy composites containing polymer modified hexagonal boron nitride. Compos A Appl Sci Manuf 107:657–664
23. Pan D, Zhang X, Yang G, Shang Y, Su F, Hu Q, ... Guo Z (2020) Thermally conductive anticorrosive epoxy nanocomposites with tannic acid-modified boron nitride nanosheets. Indus Eng Chem Res 59(46):20371–20381

24. Kim JM, Jung DW, Kim LS, Kim M, Jeong S, Lee S, ... Lee G (2021) Continuously thermal conductive pathway of bidisperse boron nitride fillers in epoxy composite for highly efficient heat dissipation. Mater Today Commun 27:102230
25. Li TL, Hsu SLC (2010) Enhanced thermal conductivity of polyimide films via a hybrid of micro-and nano-sized boron nitride. J Phys Chem B 114(20):6825–6829
26. Yi W, Wu W, Drummer D, Shen W, Chao L, Ning W (2020) Synergistic construction of thermally conductive network in polybenzoxazine with boron nitride hybrid fillers. Mater Res Express 6(12):125369
27. Fang H, Bai SL, Wong CP (2017) Thermal, mechanical and dielectric properties of flexible BN foam and BN nanosheets reinforced polymer composites for electronic packaging application. Compos A Appl Sci Manuf 100:71–80
28. Liu B, Li Y, Fei T, Han S, Xia C, Shan Z, Jiang J (2020) Highly thermally conductive polystyrene/polypropylene/boron nitride composites with 3D segregated structure prepared by solution-mixing and hot-pressing method. Chem Eng J 385:123829
29. Liang D, Ren P, Ren F, Jin Y, Wang J, Feng C, Duan Q (2020) Synergetic enhancement of thermal conductivity by constructing BN and AlN hybrid network in epoxy matrix. J Polym Res 27(8):1–12
30. Guo Y, Cao C, Luo F, Huang B, Xiao L, Qian Q, Chen Q (2019) Largely enhanced thermal conductivity and thermal stability of ultra-high molecular weight polyethylene composites via BN/CNT synergy. RSC Adv 9(70):40800–40809
31. Feng Y, Han G, Wang B, Zhou X, Ma J, Ye Y, ... Xie X (2020) Multiple synergistic effects of graphene-based hybrid and hexagonal born nitride in enhancing thermal conductivity and flame retardancy of epoxy. Chem Eng J 379:122402
32. Jin X, Wang J, Dai L, Wang W, Wu H (2019) Largely enhanced thermal conductive, dielectric, mechanical and anti-dripping performance in polycarbonate/boron nitride composites with graphene nanoplatelet and carbon nanotube. Compos Sci Technol 184:107862
33. Zhang X, Shen L, Wu H, Guo S (2013) Enhanced thermally conductivity and mechanical properties of polyethylene (PE)/boron nitride (BN) composites through multistage stretching extrusion. Compos Sci Technol 89:24–28
34. Wang X, Wu P (2017) Preparation of highly thermally conductive polymer composite at low filler content via a self-assembly process between polystyrene microspheres and boron nitride nanosheets. ACS Appl Mater Interfaces 9(23):19934–19944
35. Sun N, Sun J, Zeng X, Chen P, Qian J, Xia R, Sun R (2018) Hot-pressing induced orientation of boron nitride in polycarbonate composites with enhanced thermal conductivity. Compos A Appl Sci Manuf 110:45–52
36. Cao L, Wang J, Dong J, Zhao X, Li HB, Zhang Q (2020) Preparation of highly thermally conductive and electrically insulating PI/BNNSs nanocomposites by hot-pressing self-assembled PI/BNNSs microspheres. Compos B Eng 188:107882
37. Chen J, Wei H, Bao H, Jiang P, Huang X (2019) Millefeuille-inspired thermally conductive polymer nanocomposites with overlapping BN nanosheets for thermal management applications. ACS Appl Mater Interfaces 11(34):31402–31410
38. Shang Y, Yang G, Su F, Feng Y, Ji Y, Liu D, ... Shen C (2020) Multilayer polyethylene/hexagonal boron nitride composites showing high neutron shielding efficiency and thermal conductivity. Compos Commun 19:147–153
39. Ou X, Chen S, Lu X, Lu Q (2021) Enhancement of thermal conductivity and dimensional stability of polyimide/boron nitride films through mechanochemistry. Compos Commun 23:100549–100557

Graphene-Analog Boron Nitride Nanomaterial and Their Photocatalytic Applications

Alimuddin and Salman A. Khan

Abstract The nanomaterials of boron nitride have attracted attention in recent years due to its properties such as high thermal stability, high surface area, excellent oxidation and corrosion resistance, low density, high mechanical strength and conductivity. There are many applications of boron nitride nanomaterials in various field such as hydrogen accumulators, luminescence, advanced insulators, functional composites, water purification and photocatalytic degradation of organic pollutants. Among these various applications photocatalysis has become a focus of research. In this chapter we have described various methods of synthesis of boron nitride nanomaterial such as arch discharge method, ball milling method, chemical vapour deposition method and Laser Ablation method. The photocatalytic applications have also been described.

Keywords Graphene · Nanomaterial · Photocatalytic application

1 Introduction

The two-dimensional material family known as boron nitride nanosheets (BNNSs) has been the subject of extensive research due to its promising properties. In 2004, after the discovery of graphene, which was the first 2D nanomaterial [1, 2], they were effectively synthesized for the first time. BNNSs are layers of atomic thickness that can be composed of up to a few factors of bulk boron nitride (BN). It is commonly believed that the thin BNNSs are chemically inert, and due to their chemical inertness and oxidation resistance, BNN might be challenging to incorporate.

Graphene analog, Boron nitride nanosheets (BNNSs) are a sheet like structure having two-dimensional morphology possessing higher chemical and thermal stability than carbon materials [3–5].

In recent years, researchers have become more interested in BNNSs as a suitable catalyst support since they can prevent supported catalyst from becoming sintered on

Alimuddin (✉) · S. A. Khan
Physical Sciences Section (Chemistry), School of Sciences, Maulana Azad National Urdu University, Hyderabad 500032, Telangana, India
e-mail: khanalim192@gmail.com

M. Jawaid and A. Khan (eds.), *Multifunctional Boron-Nitride Composites*, Composites Science and Technology, https://doi.org/10.1007/978-981-99-2866-8_5

hot spots [6, 7]. According to reports, highly hazardous intracellular reactive oxygen species could be produced by graphene-based nanomaterials [8, 9].

Graphene analogues are the hexagonal boron nitride nanosheets (h-BNNSs) [10–12], it is also known as white graphene, in which N and B atoms are present alternatively in place of C atoms, there is weak vander waals forces of attraction present between BN layer as compared to graphene. The approximate band gap was found −4.6 eV and strong ionic bond in hexagonal boron nitride nanosheets [13, 14].

The distance between carbon–carbon and boron -nitrogen in graphene and boron nitride are 1.42 A° and 1.45 A° respectively, therefore to control the formation of interfaces between various materials like boron nitride and graphene are challenging [15]. To create semiconductor thermal sensor and semiconductor diode by controlled bandwidth, a hybrid of boron nitride and graphene have been used [16]. The hybrid graphene- boron nitride nanosheets mostly used in manufacturing of transistors and field effect diodes so that boron nitride used as insulator [17, 18].

The excellent thermal conductivity, strong mechanical strength, high thermal stability, wide surface areas, excellent oxidation resistance, and good chemical inertness [19–21] of hexagonal boron nitride (BN) sheet have drawn a significant amount of attention.

BN sheet is used in a variety of applications, including catalysts [22], dielectric materials [23], sensors [24], composite materials [25], and hydrogen storage [26]. In addition to these uses, there have been investigations into the possibility of using BN materials in the context of a drug delivery system [27–29]. For instance, Gnatyuk and colleagues evaluated the use of 2D-BN nanoparticles as a weak luminous marker, a spectroscopic marker, and as a delivery system for the anticancer medication doxorubicin [30]. Recent publications include the results of theoretical research on the adsorption of amino acids [31], DNA nucleobases [32–34], drug molecules [35–37], and ionic liquids [38] on BN sheets.

BN is quite similar to carbon in terms of its crystal structure, and it can be divided into four categories: graphite-analogue hexagonal boron nitride (h-BN), diamond-analogue cubic boron nitride (c-BN), rhombohedral boron nitride (r-BN), and wurtzite boron nitride (w-BN). Both c-BN and w-BN exhibit sp^3 hybridization on the B–N bonds when they are in a low-density phase. On the other hand, h-BN and r-BN both possess a high-density phase that is characterised by B–N bonds that are sp^2 hybridised [39]. Boron nitride crystal structure are shown in Fig. 1.

Boron (B) and nitrogen (N) atoms with a 1:1 stoichiometry is arranged in a sp^2 hybridized hexagonal arrangement to form hexagonal boron nitride (h BN) monolayers. These interesting, nanostructure materials have been identified as hexagonal boron nitride monolayers (h BN). These sheets are structurally and electronically identical to graphene [40], with boron (B) and nitrogen (N) surrounding carbon (C) in the periodic table. Each boron nitride atoms contain eight valence electrons that participate in the formation of both the strong and weaker in-plane and bonds. In analogy to the carbon-based counterparts, bulk-like h BN is composed of two-dimensional (2D) h BN sheets, which have a layered structure similar to that of graphite. Additionally, BN nanotubes and structures similar to fullerenes can be seen [41, 42]. Furthermore, based on first-principles calculations [43–46], It is predicted

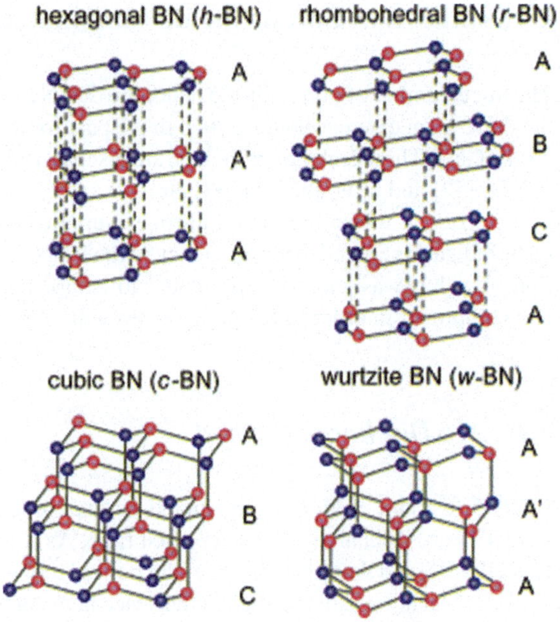

Fig. 1 Structures of four type of Boron Nitride. Reproduced from [83] Copyright 2019, Materials today, Elsevier Inc

that one-dimensional (1D) h BN nanoribbons and porous two-dimensional (2D) networks exist, such as BNyne, BNdiyne, and BNtriyne (analogues of graphyne, graphdiyne, and graphtriyne).

The binary elemental composition of BN reveals notable variations between the various allotropes of the carbon atom. Because of the disparity in the electronegativity of the two elements, h-BN undergoes a significant charge transfer from the element B to the element N, which leads to the formation of an insulator with a large band gap. This change in the material's electronic structure is responsible for the transformation. This results in an ionic quality being present. In addition to possessing a gap of approximately 6 eV [47–49], h BN also possesses a low dielectric constant. This material does not have any dangling bonds, a high breakdown voltage, a low defect density, and dangling bonds are not present. In addition to having great thermal conductivity, it possesses superior chemical and thermal resistance.

Porous boron nitride nanosheets were developed by Lei et al. by utilising the large specific surface area and high adsorption capacity of boron nitride (BN) [50]. Oil, organic solvents, and contaminants can be effectively removed from water by using a nanosheet of the type being described here. As a direct result of this, the ability of BN nanosheets to absorb substances is even superior to that of adsorbents currently available on the market and those considered to be cutting edge. As a result of their high hydrophobicity, porous BN nanosheets are capable of selectively decomposing oil, organic solvents and dyes, which makes it possible for these substances to be used. There are a variety of uses in several fields, including the cleansing and treating of water.

2 Synthesis of Boron Nitride Nanomaterials

The huge number of boron nitride nanotube (BNNT) was produced using techniques that have been extensively described in the past for their application in the manufacture of CNT. These techniques include arc discharge [51], chemical vapour deposition (CVD) [52] and laser ablation [53] etc.

The pace at which boron and nitrogen supplies are converted into boron nitride (BN) radicals is crucial to the efficiency of the BNNT synthesis process. Each fabrication procedure was scientifically created utilising different precursors, circumstances, and equipment in order to facilitate the growth of BNNT.

2.1 Arc Discharge Method

In 1995, Chopra et al. [54] conducted an experiment that led to the discovery of BNNT in arc discharge for the very first time. As a step in the process of producing BNNT, a BN rod was placed inside of a hollow tungsten anode electrode and loaded with BN. An arc plasma formed in the space between the cathode and the anode, and it swiftly vaporised the anode component. The inner diameter and length of fabricated BNNTs was 1–3 nm and 200 nm respectively with one end wrapped around a dense particle. This dense particle was most likely composed of tungsten or a mixture of tungsten, boron, and nitrogen. After that, chemical compounds containing boron that are extremely conductive (YB_6, HfB_2) [55, 56] or a mixture of boron with nickel, cobalt, and/or other metals [57] were utilised in its place, and nitrogen was used as the surrounding gas. Not only did this make the anode more conductive, but it also provided catalytic activity, which was necessary for the creation of BNNTs. Narita et. al. established the catalytic influence of yttrium on the production of few-walled BNNTs (4–10 layers) based on the presence of YB_2 particles at the ends of the nanotubes [55]. Narita et. al. obtained BNNTs by arc-melting YB_6 powder to produce them. The fact that the BNNTs had anywhere from four to 10 layers served as the foundation for this piece of evidence.

By utilizing a high-purity, hot-pressed HfB_2 electrode, Loiseau et al. [56] were able to successfully synthesize BNNT. They found a significant number of nanotubes with decreased layer counts and increased thickness at the ends. In an arc discharge, Cumings et al. [57] were able to manufacture an abundance of double-walled BNNTs by using a conducting electrode that was rich in boron and adding a negligible quantity of nickel and cobalt as major catalysts. This allowed them to accomplish their goal. The introduction of the catalysts led to the successful completion of this task. The outcomes of a study that Yeh and his colleagues carried out indicate that the selection of cathode material can be flexible independent of the material that is utilized for the anode in the device [58].

When tungsten-rich cathodes or boron-rich cathodes were arced with an anode, the author made the discovery that the grade of BNNTs produced was comparable.

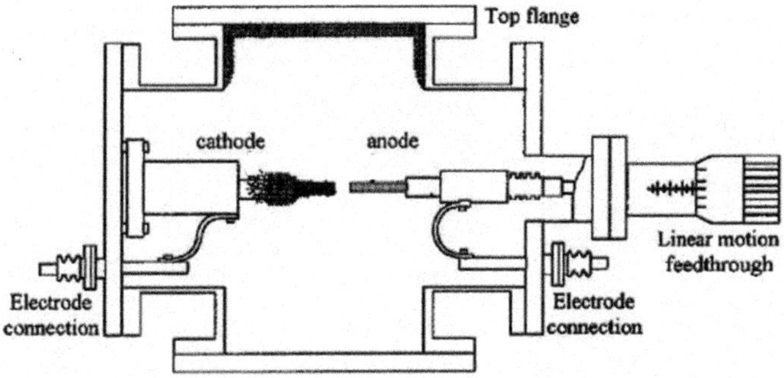

Fig. 2 Schematic diagram of Arc discharge method. Reproduced from [84] Copyright 1996, Journal of Applied Physics, AIP Publishing

Despite this, it is possible to enhance the stability of arc plasma by using tungsten or another refractory metal that has a high melting point. Because of the extremely limited space available in the reaction zone at the arc core, the production of BNNT on a large scale utilizing the arc discharge process is extremely challenging (Fig. 2).

2.2 *Ball Milling Method*

Ball milling is a potentially useful method for reducing the expense and time required to produce BNNT on an industrial scale. The direct interaction between boron and nitrogen under natural conditions can, in general, be sped up by the inclusion of a flawed or amorphous structure in the powders containing the starting material for boron.

This modification in perspective is simple to achieve by delivering a sufficient amount of mechanical energy, the control of which is determined by a number of parameters including milling time, as well as milling intensity (round per minutes). Because of this, an extremely high quantity of BNNT may be manufactured. in a normal course of events. Milling is an important phase in the process that we are going through. The treatment of boron powder, followed by annealing, has an effect on the most important part in the production of BNNT. This treatment can take a significant amount of time, which can be extended to dozens or even hundreds of hours. In an experiment that was reported on by Chen et al., boron powder was used. The powder was ball milled for a total of 150 h in NH_3 gas, and then treated to isothermal annealing at temperatures ranging from 1000 to 1200 °C [59].

In a follow-up article, the authors proposed that increasing the milling time would be the most important factor in achieving a high yield of BNNT because it would promote the nitration process between boron and NH_3, thereby forming more and more nucleation structures that aid in the formation of BNNT [60]. Because more nucleation structures could form with longer milling times, this was the case. According to these studies, 1200 °C is the optimal annealing temperature. We settled

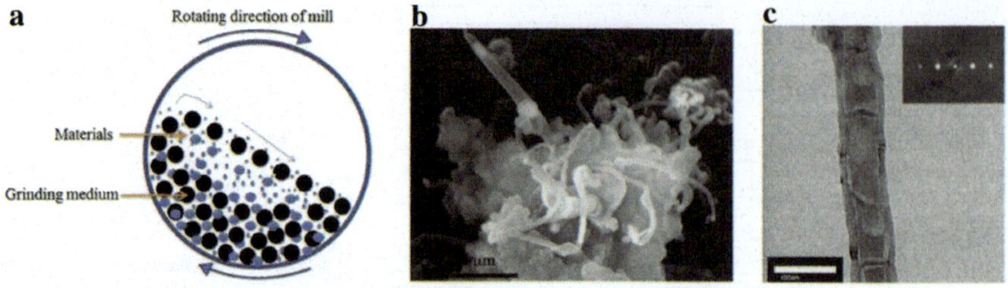

Fig. 3 **a** Diagram of ball milling process **b** SEM Images **c** TEM Images of Synthesized BNNTs. Reproduced from [85] Copyright 2020, Acapulco, JAI and [59]. Copyright 1999, Chemical Physics, Letters, Elsevier Science B.V

on this temperature range so that nitration by-products could be converted into BN tubular fibres as quickly as possible. Size and quantity of nanotubes were also significantly affected by milling and heating conditions [59].

For instance, you may obtain a lot higher BNNTs if you mill the materials for more and more. Furthermore, extremely thin nanotubes were formed when the material was annealed at a temperature of 1000 °C (25–50 nm). In this procedure, the nitriding reaction can be accelerated by using iron particles dispersed in boron powder as a catalyst. These iron fragments originate from both the milling ball and the storage container. "Kim et al., who studied the formation of BNNT, found that amorphous boron coated on the surface of the catalytic iron particle resulted from a stainless-steel container." They claimed that the initial shape of BN clusters on Fe seed particles controlled the type of as grown BNNT [61].

The crystalline boron powder that he used in his experiment underwent a change into an amorphous structure, which then coated the surface Fe particles and started the formation of BNNTs in the annealing phase. Due to the fact that the milling process was carried out in nitrogen gas, there was no nitriding reaction that took place. The initial state of bamboo-like nanotubes was an amorphous nano shell that was located on the surface of spherical iron particles. Opened nano of purity since, in addition to BNNT, it typically consists of impurities such as amorphous boron and synthetic boron nitride.cylindrical BN clusters were the cause of the growth of cylindrical nanotubes. In the annealing process, the use of ammonia rather than nitrogen results in the production of BNNTs that have a diameter of just a few nanometers and a exceptionally cylindrical structure [62]. It is important to note that the final product has a relatively low level (Fig. 3).

2.3 *Chemical Vapour Deposition Method*

Chemical vapour deposition (CVD) is the technique that is utilized the most frequently in the manufacture of carbon nanomaterials. Subsequently, this technique was applied in the synthesis of boron nitride (BN) nanomaterials. This technique has superior controllability of growth parameters than other methods, such as precursors, temperature, growth mechanism, catalysts, and experimental setup [63], which allows for the production of high-quality nanomaterials. Using the chemical vapour deposition method, Lourie et al. successfully synthesized BNNT by using borazine obtained from the chemical reaction between ammonium sulfate and sodium tetrahydridoborate ($NaBH_4$) as well as a catalyst consisting of NiB, Co, Ni_2N, and Ni at temperatures ranging from 1000 to 11,000 °C [64].

Wood et al. [65] synthesized hexagonal BN spheres by Chemical vapour deposition (CVD) in two steps. Trimethyl borate $[B(OMe)_3]$ and ammonia precursors were used in the synthesis of B-N-O particles. After that, a high temperature ammonia atmosphere was used to drive off the oxygen molecules in the spheres. After eliminating oxygen $(OMe)_3$ BNH_3 complex was generated in place of oxygen in the second phase [66].

In a different investigation that was described by Kim et al., a purified borazine source was utilized in combination with the floating catalyst nickelocene. This has been done to avoid the generation of unwanted BN species or the possibility of chemical contamination (Fig. 4a) [67, 68]. It was reported in detail how the experiment was set up, in which there are three different gas streams:

(1) NH_3 containing nickelocene acting as the catalyst (2) Borazine, which has N_2 gas within it, is introduced into the hot zone of the heating device. The formation of BNNT was under the control of flow ratio of ammonia to nitrogen at a temperature of 1200 °C, the ratio of ammonia to nitrogen is 100:3 sccm. As these produced BNNTs that have been obtained as a result of this technique were composed primarily of the double-walled nanotubes with a 2 nm diameter and length on the order of many hundreds of nanometers (Fig. 4b).

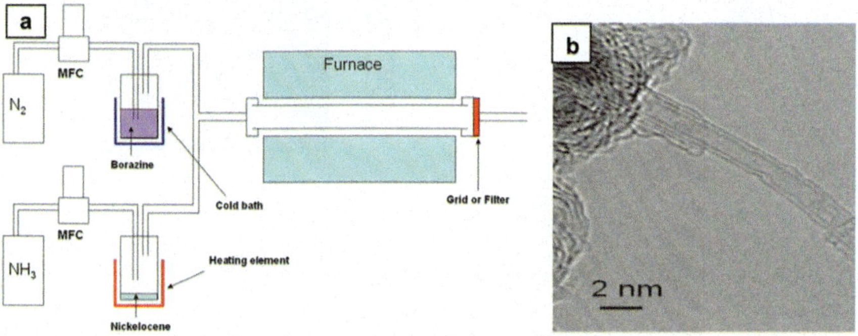

Fig. 4 **a** Schematic diagram of catalytic vapour deposition **b** Images of TEM of the double-walled BNNT. Reproduced from [68]. Copyright, 2018, Nano Convergence, Springer Open

2.4 Laser Ablation Method

Laser Ablation is a process in which CO_2 laser passed in an inert gas filled chamber which vaporises Boron or BN source. The vaporized boron is being allowed to nucleate and develop [69]. The nanotubes are being deposited on the cooler surfaces of the reactor. Collectors are those who deployed at varying heights and places across the system in order to collect nanotubes [70, 71].

Laser ablation method was introduced by Golberg et.al. for the synthesis of BNNTs [72]. He developed MWBNNTs by introducing a crystal of cubic BN to a CO_2 laser that was heated to 4727 °C while under high nitrogen pressure of 5–15 gigapascals. In further study, a hexagonal BN target [73] was utilized as a precursor in order to produce highly crystalline BNNTs without the use of a catalyst [74]. These BNNTs had a minimal number of defects and few wall structures overall. The final BNNT product was found to contain impurities such as amorphous boron fakes, BN cones, and BN onions. These impurities were discovered during quality control testing.

BNNT was synthesized by Kim et al. [69] by using boronfibers with a diameter of 102 μm and nitrogen gas as precursors in a atmosphere of high pressure and temperature (HPT) laser ablation. Boron fiber was heated at 1000 watt with CO_2 laser in nitrogen gas. At a point of irradiation, the temperature reaches upto 1501 K. BNNTs which was cotton like were collected in a chamber at specific height above the molten boron ball. It contains h- BN and boron nanoparticles as impurities (Fig. 5).

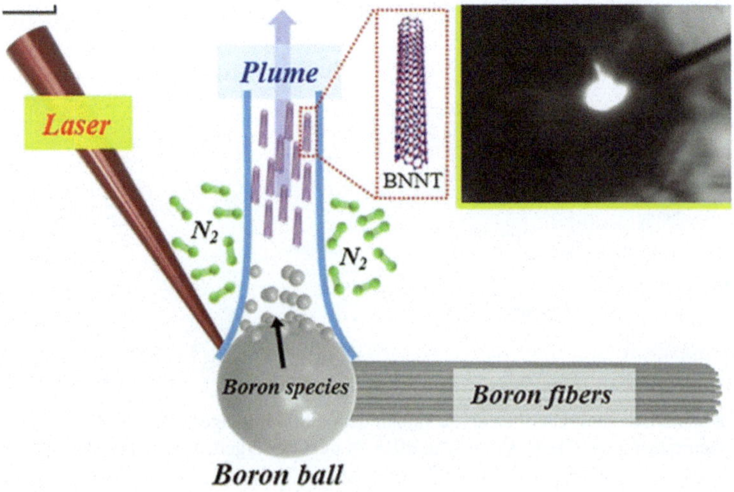

Fig. 5 Schematic diagram of BNNT synthesis in the HPT Laser ablation Method. Reproduced from [70]. Copyright, 2019, Scientific reports, Nature Portfolio

3 Photocatalytic Application of Boron Nitride Nanomaterial

The oxidation of organic compounds is a common application of photocatalysis, which is a process that is driven by light energy. The organic molecules break down into reactive free radicals, which in turn leads to the breakdown of the material. Organic compounds consisting of BN-based compounds have shown to be effective in significant potential for the photocatalytic elimination of pollutants such as heavy metals, organic molecules, dyes, and pharmaceuticals.

Using a synthetic wool dyeing process, TiO_2/porous BN nanosheet composites were able to remove both Cr (VI) and Lanaset Red 2 (LR2B) sequentially [75]. Because they make it easier for electrons and holes to be separated on the TiO_2/P-BNNS, the BNNSs considerably improve the photocatalytic activity of the porous TiO_2 nanostructures. The presence of photoreaction species such as hydroxyl and hydroxyl radicals on the surfaces of TiO_2 and the wide surface area of BNNSs both contribute to the rapid adsorption of LR2B onto porous TiO_2 and BNNSs. Not only does the separation of holes and electrons result in the reduction of Cr (VI) into Cr (III), but it also results in the generation of OH, which oxidizes LR2B into Carbon dioxide and water. The composite was recycled for a total of five separate cycles in order to remove LR2B and Cr (VI) simultaneously.

The photodegradation of MB and MO was found to be possible with a BN-doped polyaniline (PANI) hybrid nanocomposite, with a 93% and 95% degradation efficiency, respectively, in 1.5 h [76, 77]. The photocatalytic efficiency of the nanocomposite is improved by doping PANI with BN, which increases the generation of photoelectrons and holes. Specifically, PANI has the conducting polymeric chains required for the breakdown of organic materials. As can be seen in Fig. 6, the photocatalytic effectiveness of all BN composites with varying loadings of BN was higher than that of bare PANI. The photocatalytic efficiency, as reported by Shahabuddin et al. [76], rose at low h-BN loadings but declined at high h-BN loadings. This was likely due to the fact that photocatalytic activity was hampered due to the aggregation of nanosheets, which decreased the surface area available for photosynthesis.

TiO_2/BN composites were among the most attracting metal oxide/BN semiconductor systems. TiO2/BN composite has been shown to exhibit a high photocatalytic performance for the degradation of methylene blue (MB), methyl orange (MO), and rhodamine B (RhB). The mechanism by which BN in the composites increases adsorption and promotes charge separation for contaminants is described. Enhanced visible light photocatalytic activity for MB degradation in water was found for BN-TiO_2 by Singh et. al. [78, 79]. With a photodegradation rate of 79% after 200 min in the visible range, BN-TiO_2 leads TiO_2 (32%). Since - conjugation between MB and BN allows MB molecules to be transported from solution to the surface of TiO_2 and adsorb with offset face-to-face orientation, the adsorption ability of dyes is enhanced in comparison to that of pure TiO_2. Under UV light irradiation. Fu et al. [80] investigated the efficacy of h-BN/TiO_2 for the degradation of RhB and MB. Compared to pure TiO2, the optimized h-BN/TiO_2 composite accelerated the photocatalytic

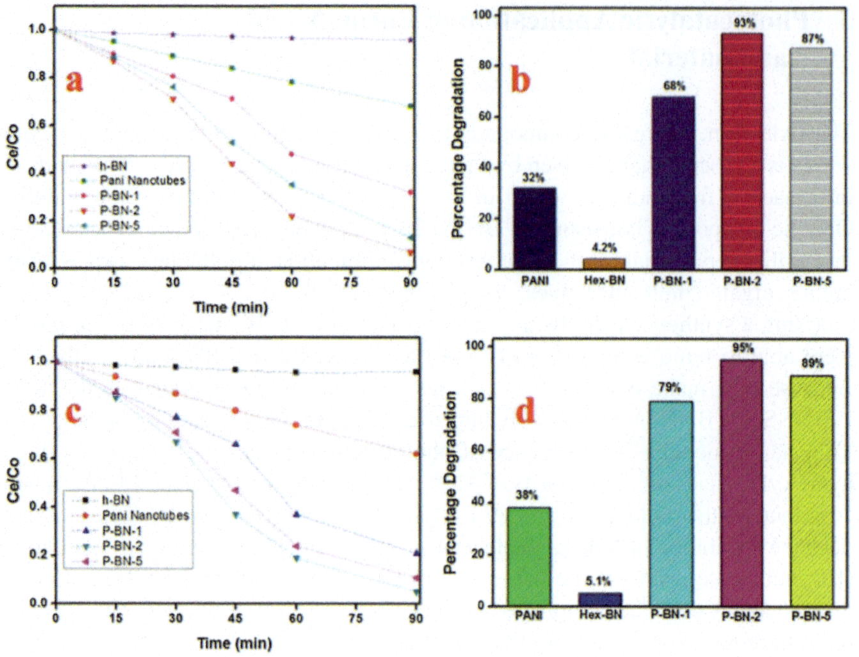

Fig. 6 **a**, **c** Rate of photodegradation of MO and MB by various photocatalysts at different time interval **b**, **d** percentage of MO and MB degradation in the presence of various photocatalysts. Reproduced from [76]. Copyright (2018), Arabian Journal of Chemistry, The Authors, Elsevier B.V. and King Saud University

degradation of RhB and MB by a factor of 15 and 8, respectively. Electrostatic repulsion brought together BN nanofiber and TiO_2. Under UV light, the composite successfully catalysed the decomposition of MO [81]. The composite's removal rate constant was determined to be optimal.

The photocatalytic activity of the AgI-BN nanocomposites for RhB and anti-photo corrosion was found to be much higher than that of pure AgI when the conditions were simulated to be similar to those of solar light irradiation [77, 82]. The photocatalytic activity of AgI was significantly improved by hybridization with BN nanosheets, which is mostly attributed to the layered structure, hydrophobic nature, and high surface area of BN. After seventy minutes, the AgI-BN nanocomposite had decomposed approximately 96.17% of the RhB. In the presence of $NaBH_4$, the composite was also effective in converting 4-nitrophenol to 4-aminophenol. Figure 7 illustrates the photocatalytic performance of AgI-BN nanocomposites with a variety of BN loadings [82].

Following the completion of the dye adsorption process, the composite was filtered, rinsed with water, then dried before being recycled for another one of the five cycles. The level of degradation that is achieved drops with each cycle, and the most recent cycle achieved only 25% of the RhB degradation that was possible. The

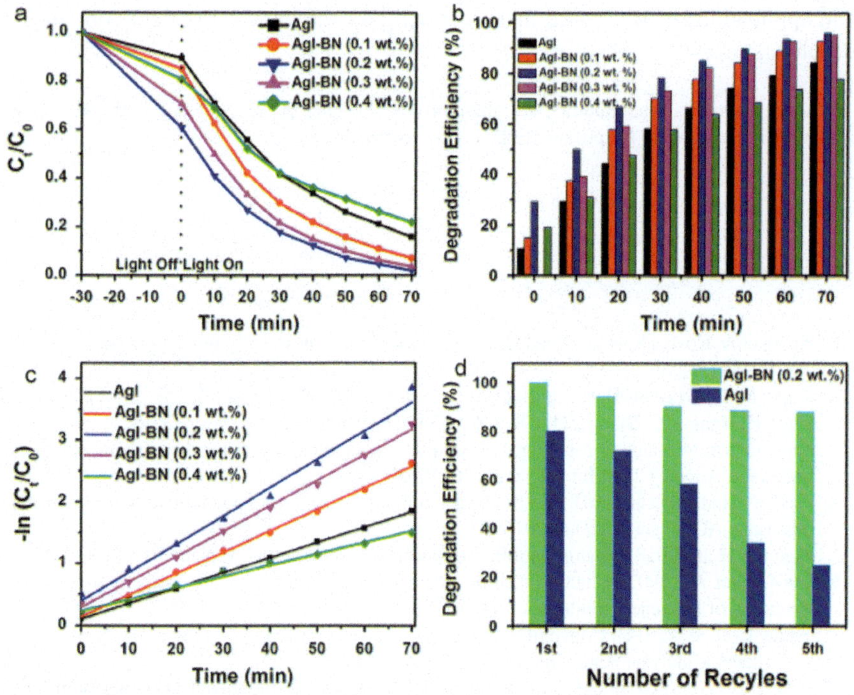

Fig. 7 **a** Photocatalytic degradation of RhB by simulated sunlight-irradiated AgI and AgI–BN nanocomposites. **b** RhB degradation by AgI and AgI–BN nanocomposites over time **c** plot of -ln C_t /C_0 versus solar irradiation time for photocatalysis of RhB solution with AgI and AgI–BN nanocomposites. **d** reusability of the AgI–BN (0.2 wt%) nanocomposite during RhB solution degradation by solar radiation. Reproduced from [82]. Copyright, 2015, Ceramics International, Elsevier Ltd and Techna Group S. r.l

RhB is oxidized, which results in the formation of mineralized chemicals, water, and carbon dioxide.

4 Conclusion

In photocatalysis, boron nanomaterials prove exclusive physiochemical characteristics and excellent photocatalytic activity was shown by the photocatalyst found by combining with other semiconductors. To address the photocatalytic defects a summary of modification scheme was described in this chapter. Though poor stability, high electron–hole pair recombination rate and large forbidden bandwidth are the main challenges in its practical applications. As research continues to be conducted in this area, it is anticipated that the technological hurdles connected to the synthesis and applications of BN-based materials will be overcome as the field continues to develop.

Acknowledgements The authors are thankful to Maulana Azad National Urdu University, Gachibowli, Hyderabad, for providing the necessary facilities.

References

1. Novoselov, Kostya S, et al. (2004) Electric field effect in atomically thin carbon films. Science 306.5696:666–669
2. Corso M et al. (2004) Boron nitride nanomesh. Science 303.5655:217–220
3. Chen H, Yang ZZ, Zhang ZH, Chen ZT, Chi MF, Wang S, Fu J, Dai S (2019) A strategy for construction of a nanoporous highly crystalline hexagonal boron nitride from an amorphous precursor. Angew Chem Int Ed 58:10626–10630
4. Guiney, Linda M et al. (2018) Assessing and mitigating the hazard potential of two-dimensional materials. ACS Nano 12(7):6360–6377
5. Mateti S et al. (2018) Biocompatibility of boron nitride nanosheets. Nano Res 11(1):334–342
6. Lin C-A et al. (2002) Characterization of boron-nitride-supported Pt catalysts for the deep oxidation of benzene. J Catalysis 210(1):39–45
7. Wang L et al. (2011) Convenient synthesis and applications of gram scale boron nitride nanosheets. Catalysis Sci Technol 1(7):1119–1123
8. Bianco A (2013) Graphene: safe or toxic? The two faces of the medal. Angew Chem Int Ed 52(19):4986–4997
9. Seabra, Amedea B et al. (2014) Nanotoxicity of graphene and graphene oxide. Chem Res Toxicol 27(2):159–168
10. Golberg D et al. (2010) Boron nitride nanotubes and nanosheets. ACS Nano 4(6):2979–2993
11. Li J et al. (2017) Excellent performance for water purification achieved by activated porous boron nitride nanosheets. Mater Chem Phys 196:186–193
12. Li H et al. (2015) Reduced graphene oxide/boron nitride composite film as a novel binder-free anode for lithium ion batteries with enhanced performances. Electrochimica Acta 166:197–205
13. Zunger A, Katzir A, Halperin A (1976) Optical properties of hexagonal boron nitride. Phys Rev B 13(12):5560
14. Xue Y et al. (2013) Excellent electrical conductivity of the exfoliated and fluorinated hexagonal boron nitride nanosheets. Nanoscale Res Lett 8(1):1–7
15. Sutter P et al (2012) Interface formation in monolayer graphene-boron nitride heterostructures. Nano Lett 12(9):4869–4874
16. Eshkalak, Einalipour K, Sadeghzadeh S, Jalaly M (2020) Thermal resistance analysis of hybrid graphene-boron nitride nanosheets: the effect of geometry, temperature, size, strain and structural defects. Comput Mater Sci 174:109484
17. Liu H et al. (2016) Disparate strain dependent thermal conductivity of two-dimensional penta-structures. Nano Letters 16(6):3831–3842
18. Jeong H et al. (2015) Semiconductor–insulator–semiconductor diode consisting of monolayer MoS2, h-BN, and GaN heterostructure. Acs Nano 9(10):10032–10038
19. Dean CR et al. (2010) Boron nitride substrates for high-quality graphene electronics. Nature Nanotechnol 5(10):722–726
20. Pakdel A et al. (2011) Boron nitride nanosheet coatings with controllable water repellency. Acs Nano 5(8):6507–6515
21. Yu J et al. (2010) Vertically aligned boron nitride nanosheets: chemical vapor synthesis, ultraviolet light emission, and superhydrophobicity. ACS Nano 4(1):414–422

22. Marbaniang P et al. (2018) Nanorice-like structure of carbon-doped hexagonal boron nitride as an efficient metal-free catalyst for oxygen electroreduction. ACS Sustain Chem Eng 6(8):11115–11122
23. Hui F et al. (2016) On the use of two-dimensional hexagonal boron nitride as dielectric. Microelectron Eng 163:119–133
24. Zhang Y et al. (2018) Copper/hexagonal boron nitride nanosheet composite as an electrochemical sensor for nitrite determination. Int J Electrochem Sci 13:5995–6004
25. Ashton TS, Moore AL (2017) Foam-like hierarchical hexagonal boron nitride as a non-traditional thermal conductivity enhancer for polymer-based composite materials. Int J Heat Mass Transf 115:273–281
26. Ahmad P et al. (2017) Synthesis of multilayered hexagonal boron nitride microcrystals as a potential hydrogen storage element. Ceramics Int 43(9):7358–7361
27. Permyakova ES et al. (2017) Synthesis and characterization of folate conjugated boron nitride nanocarriers for targeted drug delivery. J Phys Chem C 121(50):28096–28105
28. Weng Q et al. (2014) Highly water-soluble, porous, and biocompatible boron nitrides for anticancer drug delivery. ACS Nano 8(6):6123–6130
29. Jedrzejczak-Silicka M et al. (2018) Hexagonal boron nitride functionalized with Au nanoparticles—properties and potential biological applications. Nanomaterials 8(8):605
30. Gnatyuk OP et al (2018) 2D-BN nanoparticles as a spectroscopic marker and drug delivery system with protection properties. RSC Adv 8(53):30404–30411
31. Singla P et al. (2016) Theoretical study of adsorption of amino acids on graphene and BN sheet in gas and aqueous phase with empirical DFT dispersion correction. Phys Chem Chem Phys 18(7):5597–5604
32. Anota EC et al. (2015) Non-covalent functionalization of hexagonal boron nitride nanosheets with guanine. J Molecular Modeling 21(8):1–6
33. Shakourian-Fard M, Heydari H, Kamath G (2017) Defect-based modulation of optoelectronic properties for biofunctionalized hexagonal boron nitride nanosheets. ChemPhysChem 18(17):2328–2335
34. Saikia N, Pandey R (2018) Polarity-induced surface recognition and self- assembly of noncanonical DNA nucleobases on h-BN monolayer. J Phys Chem C 122(7):3915–3925
35. Saikia N, Deka RC (2014) Density functional study on noncovalent functionalization of pyrazinamide chemotherapeutic with graphene and its prototypes. New J Chem 38(3):1116–1128
36. Saikia N, Seel M, Pandey R (2016) Stability and electronic properties of 2D nanomaterials conjugated with pyrazinamide chemotherapeutic: a first-principles cluster study. J Phys Chem C 120(36):20323–20332
37. Saikia U et al. (2017) Electronic properties of acetaminophen adsorbed on 2D clusters: a first principles density functional study. Chem Select 2(13):3613–3621
38. MShakourian-Fard M, Kamath G, Jamshidi Z (2014) Trends in physisorption of ionic liquids on boron-nitride sheets. J Phys Chem C 118(45):26003–26016
39. Yu S et al. (2018) Boron nitride-based materials for the removal of pollutants from aqueous solutions: a review. Chem Eng J 333:343–360
40. Geim AK, Novoselov KS (2007) The rise of graphene. Nat Mater 6(3):183–191. https://doi.org/10.1038/nmat1849
41. Pakdel A et al. (2012) Low-dimensional boron nitride nanomaterials. Mater Today 15(6):256–265
42. Bhimanapati GR, Glavin NR, Robinson JA (2016) 2D boron nitride: synthesis and applications. Semiconductors Semimetals Vol 95. Elsevier, 101–147
43. Hu T, Han Y, Dong J (2013) Edge reconstructions of hexagonal boron nitride nanoribbons: a first-principles study. Physica E 54:191–196
44. Khan AI et al. (2017) Thermal transport characterization of hexagonal boron nitride nanoribbons using molecular dynamics simulation. AIP Adv 7(10):105110
45. Li X-D, Cheng X-L (2018) Predicting the structural and electronic properties of two-dimensional single layer boron nitride sheets. Chem Phys Lett 694:102–106

46. Zhang Y et al. (2017) First-principle study of graphyne-like BN sheet: Electronic structure and optical properties. Comput Mater Sci 136:12–19
47. Watanabe K, Taniguchi T, Kanda H (2004) Direct-bandgap properties and evidence for ultraviolet lasing of hexagonal boron nitride single crystal. Nat Mater 3(6):404–409
48. Cassabois G, Valvin P, Gil B (2016) Hexagonal boron nitride is an indirect bandgap semiconductor. Nat Photonics 10(4):262–266
49. Blase X et al. (1995) Quasiparticle band structure of bulk hexagonal boron nitride and related systems. Phys Rev B 51(11):6868
50. Lei W et al. (2013) Porous boron nitride nanosheets for effective water cleaning. Nature Commun 4(1): 1–7
51. Ebbesen TW, Ajayan PM (1992) Large-scale synthesis of carbon nanotubes. Nature 358(6383):220–222
52. Endo M et al. (1993) The production and structure of pyrolytic carbon nanotubes (PCNTs). J Phys Chem Solids 54(12):1841–1848
53. Guo T et al. (1995) elf-assembly of tubular fullerenes. J Phys Chem 99(27):10694–10697
54. Chopra NG et al. (1995) Boron nitride nanotubes. Science 269(5226):966–967
55. Narita I, Oku T (2002) Synthesis of boron nitride nanotubes by using YB6 powder. Solid State Commun 122(9):465–468
56. Loiseau A et al (1996) Boron nitride nanotubes with reduced numbers of layers synthesized by arc discharge. Phys Rev Lett 76(25):4737–4740
57. Cumings J, Zettl A (2000) Mass-production of boron nitride double-wall nanotubes and nanococoons. Chem Phys Lett 316(3–4):211–216
58. Yeh Y-W et al. (2017) Stable synthesis of few-layered boron nitride nanotubes by anodic arc discharge. Scientific Reports 7(1):1–7
59. Chen Y et al. (1999) Synthesis of boron nitride nanotubes at low temperatures using reactive ball milling. Chem Phys Lett 299(3–4):260–264
60. Chen Y et al. (2002) Large-quantity production of high-yield boron nitride nanotubes. J Mater Res 17(8):1896–1899
61. Kim J et al. (2011) Synthesis and growth of boron nitride nanotubes by a ball milling–annealing process. Acta Materialia 59(7):2807–2813
62. Yu J et al. (2005) In situ formation of BN nanotubes during nitriding reactions. Chem Mater 17(20):5172–5176
63. Ahmad P et al. (2015) Synthesis of boron nitride nanotubes via chemical vapour deposition: a comprehensive review. Rsc Adv 5(44)35116–35137
64. Lourie OR et al. (2000) CVD growth of boron nitride nanotubes. Chem Mater 12(7):1808–1810
65. Emanet M et al. (2019) Synthesis, functionalization, and bioapplications of two-dimensional boron nitride nanomaterials. Frontiers Bioeng Biotechnol 7:363
66. Tang C et al. (2008) Synthetic routes and formation mechanisms of spherical boron nitride nanoparticles. Adv Functional Mater 18(22):3653–3661
67. Kim MJ et al. (2008) Double-walled boron nitride nanotubes grown by floating catalyst chemical vapor deposition. Nano Lett 8(10):3298–3302
68. Kim J et al. (2018) Boron nitride nanotubes: synthesis and applications. Nano Convergence 5(1):1–13
69. Smith MW et al. (2009) Very long single-and few-walled boron nitride nanotubes via the pressurized vapor/condenser method. Nanotechnology 20(50):505604
70. Kim JH et al. (2019) Dual growth mode of boron nitride nanotubes in high temperature pressure laser ablation. Scientific Reports 9(1):1–10
71. Maselugbo AO, Harrison HB, Alston JR (2022) Boron nitride nanotubes: A review of recent progress on purification methods and techniques. J Mater Res, 1–21
72. Golberg D et al (1996) Nanotubes in boron nitride laser heated at high pressure. Appl Phys Lett 69(14):2045–2047
73. Cau M et al (2008) LASER-based diagnostics applied to the study of BN nanotubes synthesis. J Nanosci Nanotechnol 8(11):6129–6140

74. Zhou GW, Zhang Z, Bai ZG (1999) Catalyst effects on formation of boron nitride nano-tubules synthesized by laser ablation. Solid State Commun 109(8):555–559
75. Xie W et al. (2017) Photocatalytic TiO2/porous BNNSs composites for simultaneous LR2B and Cr (VI) removal in wool dyeing bath. J Photochem Photobiol A: Chem 333:165–173.
76. Shahabuddin S et al. (2018) Synthesis of 2D boron nitride doped polyaniline hybrid nanocomposites for photocatalytic degradation of carcinogenic dyes from aqueous solution Arabian J Chem 11(6):1000–1016
77. Ihsanullah I (2021) Boron nitride-based materials for water purification: progress and outlook. Chemosphere 263:127970
78. Singh B et al. (2017) Nanostructured BN–TiO2 composite with ultra-high photocatalytic activity. New J Chem 41(20):11640–11646
79. Zhou C et al. (2018) Semiconductor/boron nitride composites: synthesis, properties, and photocatalysis applications. Appl Catalysis B: Environ 238:6–18
80. Fu X et al. (2013) Ball milled h-BN: an efficient holes transfer promoter to enhance the photocatalytic performance of TiO2. J Hazardous Mater 244:102–110
81. Nasr M et al. (2017) Enhanced photocatalytic performance of novel electrospun BN/TiO 2 composite nanofibers. New J Chem 41(1):81–89
82. Choi J, Amaranatha Reddy D, Kim TK (2015) Enhanced photocatalytic activity and anti-photocorrosion of AgI nanostructures by coupling with graphene-analogue boron nitride nanosheets. Ceramics Int 41(10):13793–13803
83. Zhang X, Meng J (2019) Recent progress of boron nitrides. Ultra-wide bandgap semiconductor materials, 347–419
84. Saito Y et al. (1996) Carbon nanocapsules and single-layered nanotubes produced with platinum-group metals (Ru, Rh, Pd, Os, Ir, Pt) by arc discharge. J Appl Phys 80(5):3062–3067
85. Acapulco JAI (2020) Synthesis of low dimensional boron nitride nanomaterials for thermal management applications. Diss. University of Oxford

An Extensive Study on Thermo-Mechanical Characterization of Hexagonal-Boron Nitride Polyester Composites

Debasmita Mishra and Anish Khan

Abstract The research aims at the manufacture and characterization techniques of hexagonal boron nitride reinforced composites using polyester matrix. In this work, the methods of testing and their results for mechanical, thermal and surface properties in polyester matrix composites having particulate BN of different proportions are conversed about. An assessment of the effective thermal conductivity was carried out and hence validation of the same was done with the help of experiments. The effect of inclusion of BN on the density, water absorption, tensile strength, compressive strength, micro-hardness, effective thermal conductivity (k_{eff}), glass transition temperature (T_g), coefficient of thermal expansion (CTE) and surface morphology are discussed. The thermal conductivity shows an escalation when conductive BN reinforcement is added to the polyester matrix. The coefficient of thermal expansion (CTE) decreases with addition of BN but the Glass transition temperature (T_g) tends to be enhanced a lot. Compared to analytical models, the proposed correlation is estimated to be a very good analytical model to estimate k_{eff} of composites within the percolation limit. Owing to low cost and higher mechanical and thermal properties, polyester composites serve as a better alternative to general epoxy composites. These highly conductive, low density and light-weight boron nitride reinforced polyester composites can be utilized in various areas of aviation, automotive and electronic encapsulation etc. Again, a hybrid composite was manufactured with a organic reinforcement which tends to show excellent properties for specific applications.

D. Mishra (✉)
Department of Mechanical Engineering, VSSUT, Burla, Odisha 768018, India
e-mail: debasmita.mishra01@gmail.com

A. Khan
Chemistry Department, Faculty of Science, King Abdulaziz University, Jeddah 21589, Saudi Arabia

M. Jawaid and A. Khan (eds.), *Multifunctional Boron-Nitride Composites*,
Composites Science and Technology, https://doi.org/10.1007/978-981-99-2866-8_6

Abbreviations

k_c	Thermal conductivity of the composite (W/mK)
k_m	Thermal conductivity of the matrix (W/mK)
k_f	Thermal conductivity of the filler (W/mK)
k_{f1}	Thermal conductivity of the first filler
k_{f2}	Thermal conductivity of the second filler
ϕ_1	Volume fraction of 1st filler
ϕ_2	Volume fraction of 2nd filler
ϕ	Volume fraction of the dispersed phase
Ψ	Volume fraction of the matrix phase
A	Constant (Shape factor for filler)
B	Constant (Shape factor for filler)
k_{eff}	Effective thermal conductivity of the composite
k_p	Thermal conductivity of the 1st filler
k_g	Thermal conductivity of the 2nd filler
ρ_m	Density of the matrix phase of the filler
ρ_p	Density of the 1st filler
ρ_g	Density of the 2nd filler
v_f	Volume fraction of the filler

1 Introduction

1.1 Introduction

Packaging in electronic equipment especially micro-circuitry is especially very important aspect due to rapid growth in electrical, electronics and digital technology. As electronic circuits are becoming more and more tiny and compact, heat entrapment in the circuits is a major issue. This heat should be driven away as soon as possible to prevent overheating. For that the materials chosen for packaging should have good thermal conductivity apart from having good mechanical and physical properties. They also should have low thermal coefficient of expansion. To have good propagation of signal, they should also have less relative permittivity. Furthermore, the advancements in the automobiles and aircrafts need type of materials that have good dissipation of heat as emitted by the combustion in the engine which also affect parts like engine hood, exhaust mufflers etc. These materials should have good durability, high strength to weight ratio and high thermal conductivity but a low thermal expansion with low density. The materials should also be able to deal with mechanical and thermal stresses generated due to various loading conditions and moving parts. For these ceramic materials like nitrides such as boron nitride (BN) and aluminum nitride (AlN) with good electrical resistivity having high electrical resistivity and thermal conductivity can be very well used in this work. The reinforcement Boron nitride has a resistivity value of 10^{14} Ω cm, a relative permittivity as low as 4.88, 53 kV/mm is the breakdown strength which is very good compared to other inorganic materials and most of the ceramics. Heat transfer rate of BN is also pretty well. Therefore, BN can serve as a very good inorganic reinforcement in electronic industry. But about the dielectric properties of BN reinforced composites there is very less published research articles [1–3].

Thermal properties are improved with the help of boron nitride particulates filled inside polyester matrix to get tailor-made composites for required applications [2, 4, 5]. These kinds of composite materials are beneficial due to their properties like resistance to wear and tear, corrosion, less density simpler in processing as well low cost. Zhou et al. [5] did research on HDPE composites by varying the dispersion rates of the reinforcements i.e. boron nitride and explained about their thermal conductivity on the thermal conductivity and temperature resistance. Wattanakul et al. [6] fabricated a hybrid of inorganic and inorganic fibers in a thermosetting resin to manufacture a new composite with good thermal characteristics. It comprised of resin epoxy with a diamine hardener along with particulate boron nitride. Hill and Supancic [7] did his research work on platelet shaped ceramic reinforcements which created conducting paths and then improved their thermal conductivity rapidly.

For lowering the weight, increasing strength and for enhancing thermal conductivity, reinforcements having high thermal conductivity are used in the matrix material according to different volume fractions for various applications. So, in this research the conductivity of the inorganic ceramic fillers as well as hybrid organic–inorganic fillers like boron nitride with peanut husk particulates are used and their thermal properties like thermal expansion coefficient, glass transition temperature, thermal conductivity etc. as well as relative permittivity and mechanical properties like tensile strength, compressive strength etc. were analysed theoretically as well as experimentally.

Research Objectives

There are a few research objectives carefully formulated after selective analysis of research gap from earlier investigations which are as under:

- Formulation of Theoretical models meant for evaluation of effective thermal conductivity of composites with one and multiple reinforcements.
- Manufacture of a series of polyester composites with BN reinforcement, with peanut husk particulate reinforcement and both BN and PH hybrid reinforcements as well as measuring their thermal conductivities.
- The comparison of results obtained from formulated proposed model with theoretical as well as validating with the experimental values.
- The Mechanical, physical and surface property analyses of the composites to estimate the life and sustainability of the composites.
- Study of the reinforcement effects on the thermal characteristics like glass transition temperature and coefficient of thermal expansion of the composites and hence estimating how they can possibly be used as a replacement of traditional

materials in micro-electronic packaging applications as well as hybrid composites in electronics substrate materials, laminates and casings in engines, fins, heat dissipative materials, fins and heat sinks etc.

2 Materials

2.1 The Matrix

Out of all the resins, the chosen resin material is isophthalic IS9 marine grade polyester owing to its excellent water-resistance and good thermal properties. It can be cured at room temperature only without use of any other solvent. It only needs its corresponding hardener i.e. methyl ethyl ketone. Polyester is again the cheapest resin available.

2.2 Reinforcement Materials

Boron nitride particles of a size of around 100 μ are used as the reinforcement material due to its high thermal conductivity and water-resistant properties. A chromium-based rock was identified to be a mineral in 2013 called Boron Nitride. It has cubic as well as hexagonal forms out of which the hexagonal form is more stable form whose melting point is very high i.e. about 2973 °C. The structure has similarity to graphite and every layer is covalently bonded, therefore this structure raises it conductivity value. Peanut-husk powder of micron size is also used as a secondary reinforcement for hybrid composites due to its bio-degradable nature, ease of availability and low cost. Peanut husk is a waste which comes out of peanut oil seed and India produces 2nd largest amount of peanut oil. This is why peanut husk was a good choice for hybrid composites as it is very low in density and has good mechanical properties as well.

3 Theoretical Models

There are few equations developed by various researchers which gives a theoretical approach regarding thermal conductivity of the composites having single as well as multiple reinforcements. They are very helpful because before experimentation, they can assess values of conductivity ranges for different combinations of reinforcements and matrices. Rule of mixture model is a very good model which uses the principle of weighted mean approach.

For **Series conduction** model of Rule of Mixture theory [1, 2]

$$\frac{1}{k_c} = \frac{1 - \phi_1 - \phi_2}{k_m} + \frac{1}{k_{f_1}} + \frac{1}{k_{f_2}} \tag{1}$$

k_{f1}, k_{f2}, k_m, k_c are thermal conductivities for reinforcement-1 and reinforcement-2,of the matrix, thermal conductivity of the composite and ϕ_1 is the volume fraction of the reinforcement-1 and ϕ_2 is the volume fraction of the reinforcement-2.

Maxwell [1, 2] gave an equation for thermal conductivity of composite. He used the theory of composites in dilute form of dispersions i.e. randomly oriented spherical particles inside the matrix in a homogenous medium which is as follows

$$k_c = k_m \left[\frac{k_f + 2k_m + 2\phi(k_f - k_m)}{k_f + 2k_m - 2\phi(k_f - k_m)} \right] \tag{2}$$

where k_c, k_m and k_f are effective thermal conductivities of the composite, conductivity of the matrix and reinforcement, ϕ is the volume fraction of the reinforcements. Equation 2 is mostly used for matrix as a cubic suspension having a single particulate reinforcement inside the cubic matrix.

Lewis and Nielsen [1, 2] developed a model by using Halpin–Tsai equation and modifying it for a system which considers particulate reinforcements, their orientation and shape and assumes the system as isotropic in nature.

$$k_c = k_m \left[\frac{1 + AB\phi}{1 - B\phi\Psi} \right] \tag{3}$$

where, $B = \left[\frac{(k_f / k_m) - 1}{(k_f / k_m) + A} \right]$ where, k_f is conductivity of reinforcement and 'ϕ' is defined as the volume fraction of the reinforcement and A and B are shape factors according to the shape of the particulates.

Formulated proposed model [8–10]

Formulated model for single reinforcement in composite

A cubic volume is considered inside which a spherical particulate reinforcement is considered to be centred. The heat transfer rate can be calculated by getting the measure of the amount of heat flow through the spherical particulate reinforcement and the cubic volume considered as the matrix and then by integrating for the whole composite. Thus, the formulation for the new model was found out as under.

$$k_{eff} = \left[1/k_p \left(1 - \frac{6v_f}{\Pi}\right)^{1/3} + 2\left(\left(k_p \left(\frac{4\Pi}{3v_f}\right)^{1/3} + \Pi\left(\frac{2v_f}{9\Pi}\right) \right)^{1/3} x \left(k_g \frac{\rho_p}{\rho_g} - k_p \right) \right)^{-1} \right]^{-1} \tag{4}$$

kp is thermal conductivity of the polyester matrix and kg is thermal conductivity of the boron nitride reinforcement, ρ_p and ρ_g are the the densities of the polyester matrix and boron nitride reinforcement and v_f the volume fraction of the reinforcement.

Thermal conductivity values of boron nitride as the reinforcement as well as BN + PH reinforcements were evaluated by using the models given by earlier researchers, the formulated proposed model and then compared to the experimentally measured values. Conductivity values are found out for polyester matrix having BN as a reinforcement and also with two reinforcements BN as well as PH by using theoretical models from literature given by previous researchers. Then theses values were analysed with the experimentally measured results. Because of more contact area between the particles in the composite as the particles create a kind of network between the particles, the amount of heat transfer increases rapidly. This deformation of BN particles within the matrix improves the thermal properties of the composites as a whole.

Debasmita Mishra and Alok Satapathy in their research [8–10] developed a novel model for effective thermal conductivity of the composite where the conductivity goes on rising with the rise in weight of the reinforcement. The conductivity value enhances with conductive reinforcements like Boron Nitride and Aluminium Nitride whereas it drops if we add insulative reinforcements like Peanut husk. Again, with hybrid reinforcements owing to being an inorganic and another organic one like BN and PH in epoxy resin, we can get desirable properties with low dielectric constant values, good thermal performance and also mechanical aspects. The research was carried out both on single and hybrid reinforcements. Thermal conduction equations were hence formulated for single (BN-epoxy composites) and hybrid reinforcements (BN + PH-epoxy composites) in epoxy resin. The novel model which was formulated was found to be very much closer to the experimental as well as numerical study.

4 Results and Discussion

4.1 Physical Characteristics

4.1.1 Density

The density of the composites keeps on increasing with increase in filler content. With the rise in BN content in polyester resin (0–35 wt %), there is a rise in density of the composite by about 1.6%, though there is also an increase in the void content from 1.5 to 8.5%. For hybrid composites, with the increase in peanut husk such that BN reinforcement is not varied, the density becomes lower by about 60%. With the decrease in density, the void content still goes on increasing from 1.5 to 11.1% (Table 1).

Table 1 Measured and Theoretical densities of the composites (Polyester filled with Boron Nitride)

Composites	Measured density (gm/cc)	Theoretical density (gm/cc)	Void fraction (%)
Neat polyester (hardened)	1.87	1.9	1.57894
Polyester + BN (5%)	1.805	1.916667	5.826086957
Polyester + BN (10%)	1.81	1.933628	6.652173913
Polyester + BN (15%)	1.815	1.950893	7.221967963
Polyester + BN (20%)	1.82	1.968468	7.796338673
Polyester + BN (25%)	1.825	1.986364	8.375286041
Polyester + BN (30%)	1.85	2.004587	8.958810069
Polyester + BN (35%)	1.9	2.023148	8.558352403

4.1.2 Hydrophilicity of Composites

In this research work the inorganic BN ceramic fiber in its particulate form alongwith organic fiber peanut husk are used as the reinforcements in polyester matrix with the aim to reduce the absorption of water by the composites and also puffing up of the composites due to entrapment of moisture inside them. By using a hydrophilic fiber like BN the composites can also be made hydrophilic without altering its good mechanical characteristics. The hydrophilic nature of the composites with incorporation of BN is shown in Fig. 1a. The composite samples were first immersed in water and the weight before and after was measured after every specific intervals upto 24 h. As there is weight gain due to absorption of moisture, the percentage gain in weight was found out by taking the difference in weight of the composites for different intervals of time. It has been observed that the BN reinforcement composite with one reinforcement i.e. BN doesn't absorb much water as compared to the hybrid composites but the water uptake goes on increasing upto 168 h. At about after 168 h there is no furthermore absorption of water. Though BN is itself moisture resistant but this is due to the fact that there are voids in the composites as the reinforcement weight content goes on increasing and hence the voids are responsible for absorbing moisture inside them in the composites.

Figure 1b–d illustrate the hydrophilic characteristics of the hybrid composites (BN + PH-polyester composites) at room temperature. The composites for a constant 10 wt% of BN and 0–25 wt% of PH with PH increasing gradually are prone to gradual absorption of water with rise in percentage weight of peanut husk reinforcement due to its organic nature. The composites of 10 wt% of BN and 5 wt% of PH have lower

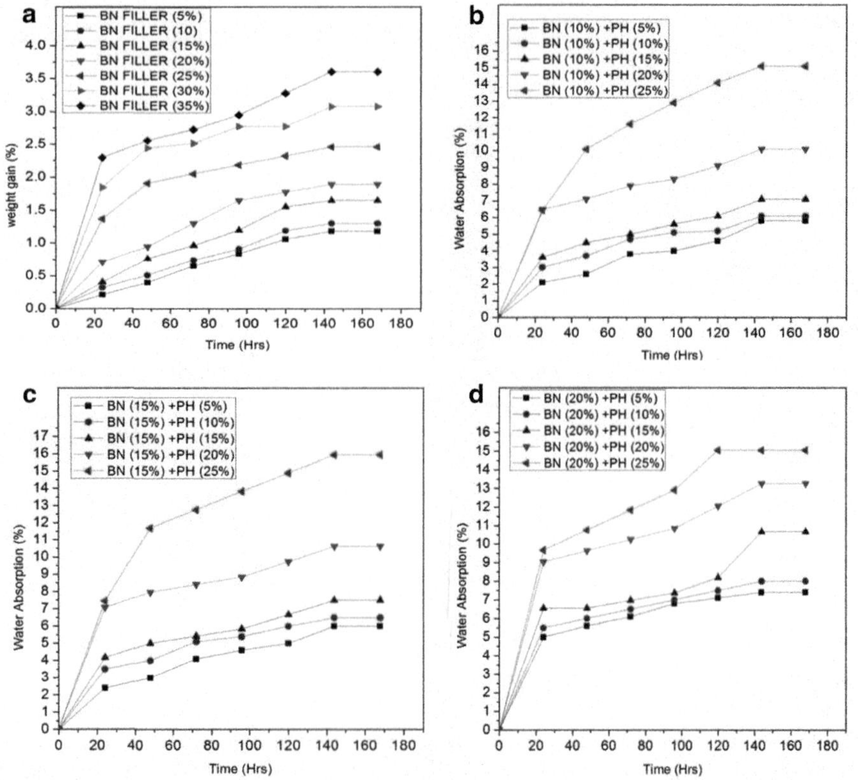

Fig. 1 **a** Immersion time on properties of single filler composites **b** Immersion Time on properties of hybrid composites. **c** Effect of immersion time on properties of hybrid composites (d) Effect of immersion time on properties of hybrid composites

absorption of moisture compared to that of samples with 10 wt% of BN and 25 wt% of PH. The composition of 15 wt% of BN and 20 wt% of PH also have less amount of water retention (Tables 2, 3 and 4).

The percentage weight of PH for 20 wt% of BN hence shows a higher absorption than 15 wt% of BN. This is because the Peanut husk is more hydrophilic in nature due to its organic nature and the voids in the composite facilitate the absorption of moisture even more whereas BN is itself moisture resistant and hence tends to impart its hydrophobic property to the composite.

So, during the processing of the composites it is very essential that the composites should have minimal voids to prevent entrapment of air and moisture by avoiding bubble formation during stirring while processing. Therefore, by optimising the effects of BN and PH the best combination of both reinforcements in the polyester matrix can be chosen which has minimal voids with least absorption of moisture with more percentages of BN with lower percentage of PH. From the figures it can be concluded that 20 wt% of BN and 5–10 wt% of PH is a very good combination

Table 2 Measured and Theoretical densities of the composites (Polyester filled with Boron Nitride (10%) + peanut husk (PH))

Composites	Measured density (gm/cc)	Theoretical density (gm/cc)	Void fraction
Neat polyester (hardened)	1.87	1.9	1.57894
Polyester + BN (10%) + PH(5%)	1.41	1.408163	0.130454
Polyester + BN (10%) + PH(10%)	1.12	1.107264	1.150223
Polyester + BN (10%) + PH(15%)	0.97	0.912317	6.322693
Polyester + BN (10%) + PH(20%)	0.84	0.77574	8.283703
Polyester + BN (10%) + PH(25%)	0.75	0.67473	11.15557

Table 3 Measured and theoretical densities of the composites (Polyester filled with Boron Nitride (15%) + peanut husk (PH))

Composites	Measured density (gm/cc)	Theoretical density (gm/cc)	Void fraction
Neat Polyester (hardened)	1.87	1.9	1.57894
Polyester + BN(15%) + PH(5%)	1.45	1.417297	2.30742
Polyester + BN(15%) + PH(10%)	1.17	1.112903	5.130456
Polyester + BN(15%) + PH(15%)	0.99	0.916143	8.061733
Polyester + BN(15%) + PH(20%)	0.85	0.778504	9.183768
Polyester + BN(15%) + PH(25%)	0.76	0.67682	12.28983

which has high strength, high strength to weight ratio, good thermal properties and minimal moisture absorption levels.

4.2 Mechanical Property Characterization

4.2.1 Tensile and Compressive Test

When the Boron Nitride is used as the reinforcement for BN-polyester composites the Tensile strength declines from about 75 MPa to about 42 MPa as the percentage

Table 4 Measured and Theoretical densities of the composites (Polyester filled with Boron Nitride (20%) + peanut husk (PH))

Composites	Measured density (gm/cc)	Theoretical density (gm/cc)	Void fraction
Neat polyester (hardened)	1.87	1.9	1.57894
Polyester + BN(20%) + PH(5%)	1.51	1.426551	5.849703
Polyester + BN(20%) + PH(10%)	1.25	1.118601	11.74673
Polyester + BN(20%) + PH(15%)	1.05	0.92	14.13043
Polyester + BN(20%) + PH(20%)	0.9	0.781287	15.19454
Polyester + BN(20%) + PH(25%)	0.79	0.678923	16.36077

weight of BN is increased from 0–35 wt%.For BN-PH-polyester hybrid composites though the tensile strength rises due to higher strength of PH fibres which is organic in nature whereas BN due to its particulate form exhibits brittle nature in the composites. The tensile strength increments to about 18.7% for 20 wt% BN and 25 wt% PH in polyester matrix. The compressive strength increases in both single as well as hybrid composites. Polyester has a compressive strength of 120 MPa and the increment is about 15.27% from 120MPA to 138 MPa when 35 wt% BN particulates are added to the polyester matrix. For hybrid composites the increment of compressive strength is high as both the fillers have high compressive strength and when the compression test is done the process fills up voids in the material which again improves the strength. Figure 2a–d shows the Tensile and compressive strength with respect to the rise in weight fraction of the reinforcements in both single as well as hybrid composites. The tests were carried out using Universal Testing Machine Instron 3382.

4.2.2 Micro-hardness

Hardness is very important property of any material especially a composite material. It is the ability to withstand indentation. It was realised that when BN was added to the matrix, then the micro-hardness is found to rise and becomes a function of the BN content. With 35 vol% of BN inclusion, the hardness is improved by about 1.78 times compared to neat polyester. The Fig. 3a, b shows the micro-hardness characteristics of samples of single as well as hybrid reinforced composites i.e. BN-polyester and BN-PH-polyester composites. The rise in hardness is higher in single reinforced composites than in hybrid composites though the difference is not so high. The micro-harness decreases slightly when PH is more in the matrix. Therefore, a compromise between the two reinforcements has to be made such that with the optimal combination of both we get the best of both with best of properties.20 wt%

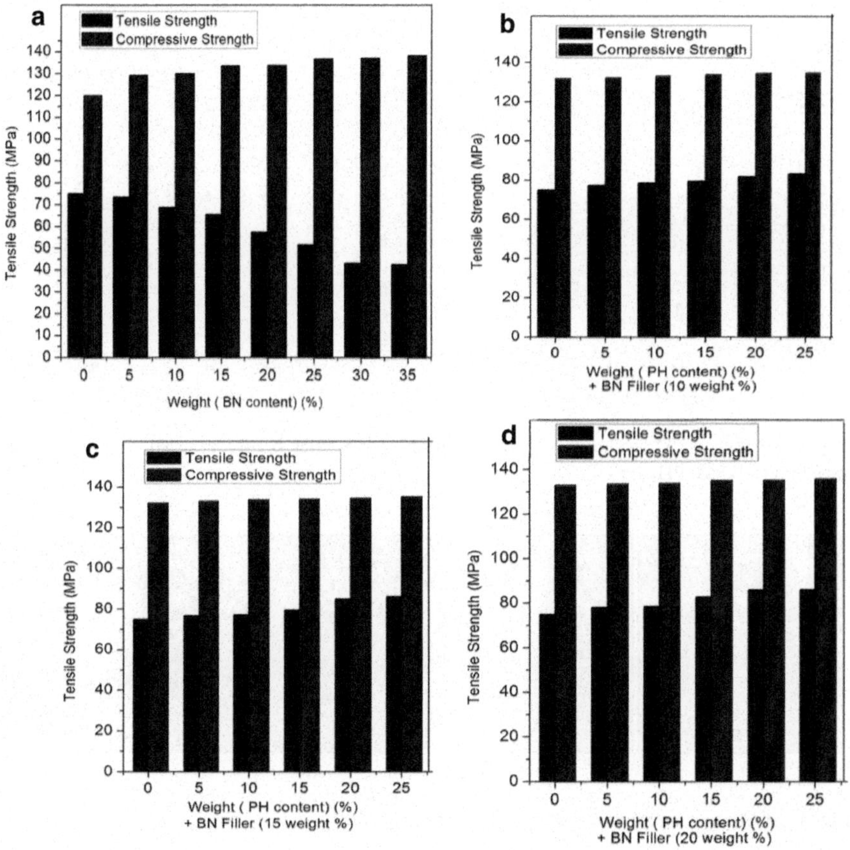

Fig. 2 **a** Effect of BN Filler addition on Tensile and Compressive strengths of the composites. **b** Effect of BN Filler addition on Tensile and Compressive strengths of hybrid filler (BN 10% + PH 25%) composites. **c** Effect of BN Filler on Tensile and Compressive strengths of hybrid filler (BN (15%) + PH 25%) composites. **d** Effect of BN Filler on Tensile and Compressive strengths of hybrid filler (BN(20%) + PH 25%) composites

with less wt% of PH is seen to give best results. As Boron Nitride is a ceramic material which has high hardness value than Peanut husk, so it imparts more hardness to the composites.

4.3 Surface Morphology

Figure 4a–d depicts the micro-structural morphology of Boron-nitride reinforced polyester composites. The particle distribution was observed to be uniform in the matrix material which is evident from the fractured surface seen in Fig. 1b. It can

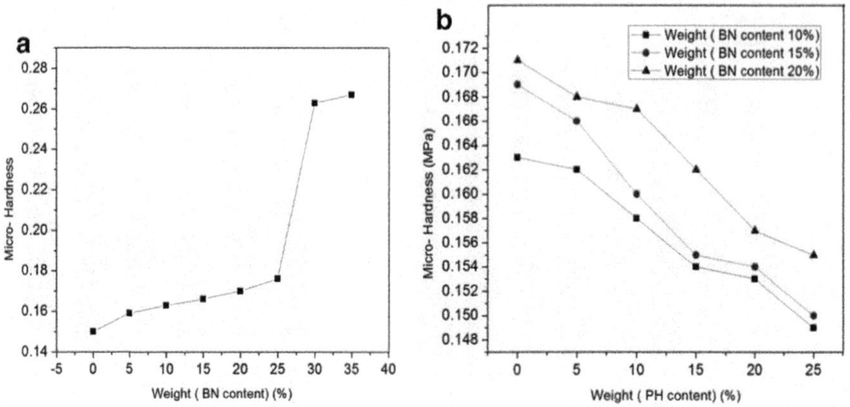

Fig. 3 **a** Effect on Micro-hardness with variation of BN filler **b** Effect on Micro-hardness with variation of BN & PH

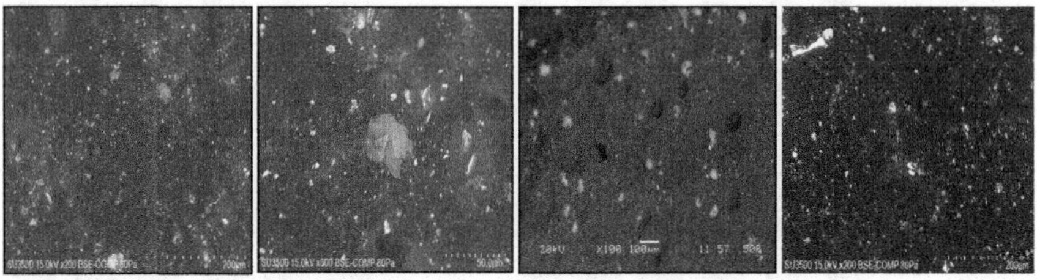

(a) (b) (c) (d)

Fig. 4 SEM images of particle-filled polyester composites

also be noticed with magnifying that there are voids due to clustering effect of the reinforcements because of mixing and curing defects discerned from Fig. 1c. Again, Fig. 1d distinguishes the surface microstructure of hybrid composites with BN and PH as reinforcements.

4.4 Thermal Characterization

4.4.1 *Effective Thermal Conductivity (k_{eff})*

The analytical models already formulated by previous researchers have been taken as the basis to calculate the theoretical values of thermal conductivities of BN-polyester composites as well as BN-PH-polyester composites. The experimental values are then compared with these evaluated values as well as results from newly formulated proposed model. Due to deformation of particles of BN within the matrix the area increases and their contact area increases due to which conductivity rises

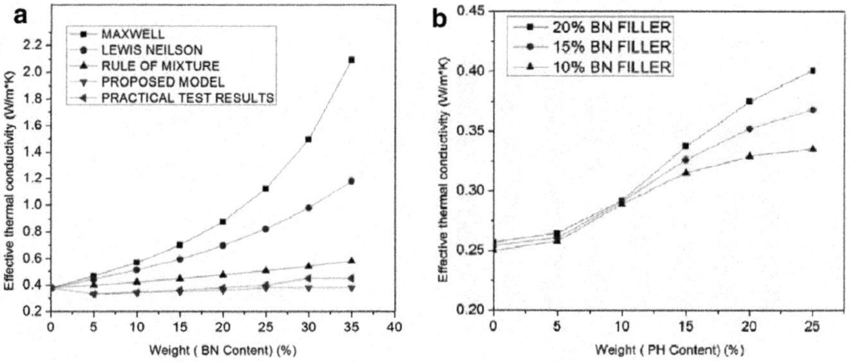

Fig. 5 **a** Comparison of k_{eff} values obtained from different models with experimental values for single filler composite. **b** Comparison of thermal conductivity values obtained from experimental estimations for hybrid composites

as the particles start forming chain-like network within themselves which escalates the conductivity value suddenly beyond the percolation limit which explains the advantage of BN particles over other particles.

In the Fig. 5a, it is shown that as the weight fraction of BN rises, the thermal conductivity (effective) (k_{eff}) also escalates. It is evident from the figure that the calculated values of the thermal conductivity using the formulated model clearly gives a good prediction of effective thermal conductivity when compared with other analytical models. Though the formulated model is a very good predictive model compared to the measured values but shows lower thermal conductivity values than the actual values. The increase in thermal conductivity is due to the highly conductive BN filler embedded in the insulative matrix which enhances its thermal properties.

The above graph (Fig. 5b) shows the variation of effective thermal conductivity with the increasing PH filler from 0 to 25 wt% keeping BN weight % fixed at particular weight fractions (10,15 and & 20 wt%) for individual sets. The above data interprets that with increase in PH weight%, for a particular weight % of BN (let say BN 10%) the thermal conductivity doesn't increase much in case of measured values. It even interprets that for the other sets of data with the increase of BN (BN 15% and 20%) k_{eff} increases by lower marginal extent for the corresponding similar PH percentages. The negligible drop in thermal conductivity value is due to PH reinforcement as it is insulative in nature and its conductivity is low but as we go on adding more BN (20 wt%), the property of PH is engulfed by the property of BN and the rise in thermal conductivity value is more or less equal to the conductivity obtained by BN reinforcement only (BN 25 wt%). Though the formulated model is in closer proximity with the measured results for reinforcements with single filler, for hybrid arrangement it is not the case.

So, for a little compromise in thermal conductivity, a new hybrid composite can be produced which consists of both an inorganic and a organic reinforcement.

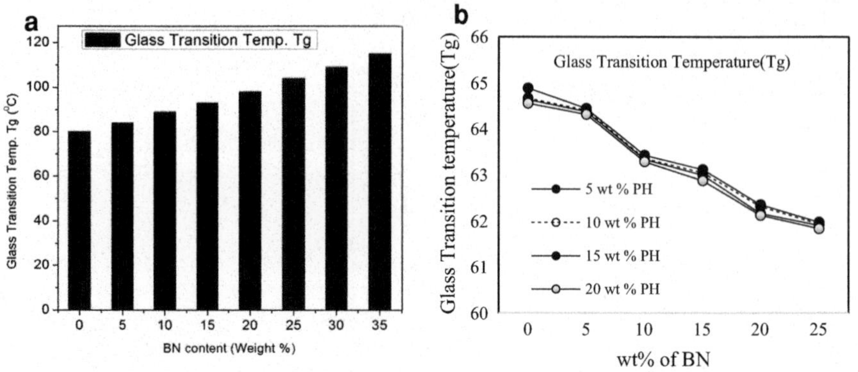

Fig. 6 **a** Glass transition temperatures of Polyester-BN composites **b** Glass transition temperatures of hybrid composites

4.4.2 Glass Transition Temperature (Tg)

The Fig. 6a below exhibits the glass transition temperatures of polyester reinforced boron nitride reinforcement. Polyester in its neat form has a glass transition temperature of about 80 °C.When the BN reinforcement is added to the polyester matrix the glass transition temperature shows a increasing trend upto 115 °C when the reinforcement goes on increasing from 0 to 35 wt%. For hybrid composites, with addition of peanut husk from 0 to 20 wt% for varying BN content from 0 to 25 wt% there a slight decrease in Tg due to insulative nature and low thermal properties of peanut husk but the decrement in comparison to polyester-BN single filler composites is as low as 2.9% which is negligible as compared to significant improvement in other properties shown in Fig. 4b.

4.4.3 Co-efficient of Thermal Expansion (CTE)

Figure 7a illustrates the co-efficient of thermal expansion of BN reinforced polyester composites with the rise in content of the reinforcement. The CTE value descends as we go on increasing BN reinforcement. It shows that the CTE of the composite gradually decreases with the increase in filler content. Polyester has a CTE value of about 65 ppm/°C which goes on decreasing to 62.157 ppm/°C with 0–35 wt% of BN, the reason being BN has less CTE value of 36 ppm/°C. The CTE value decreases with increase in peanut husk as peanut husk has itself a even lower CTE value (Fig. 7b).

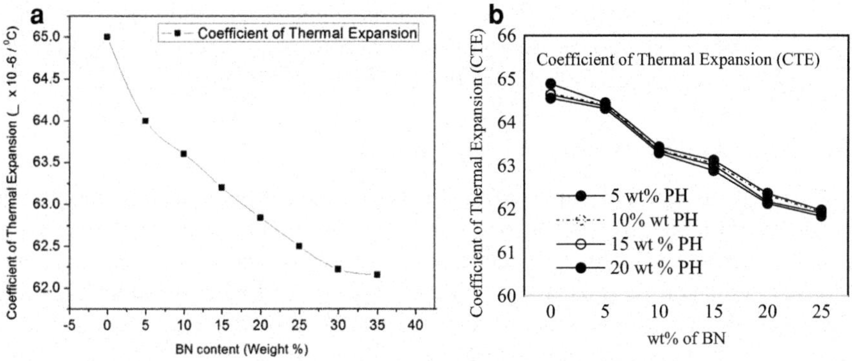

Fig. 7 **a** Variation of Coefficient of Thermal Expansion of polyester-BN composites **b** Variation of CTE of hybrids

5 Summary and Conclusion of Research Findings

1. The production of a novel sets of particulates filled polyester-BN composites and hybrid (polyester/BN/PH) composites is possible.
2. The reinforcements in the matrix modify the mechanical properties like micro-hardness as well as strength of the composites. It is observed that with the inclusion of BN, tensile strength decreases in the composites and this decrement is due to the nature of particulate fibers. Neat polyester has a tensile strength of 75 Mpa which decreases to 42.50 MPa with the addition of 35 wt% of BN in it. In hybrid composites rather it increases due to presence of peanut husk which helps in imparting strength to the composites.
3. The physical properties like water absorption behaviour of these composites is also highly influenced by the type and content of fillers.
4. It was observed that the not all empirical models serve as very good models for particulate inclusions and that Rule of mixture correlation as well as the proposed correlation can be used to estimate k_{eff} for composites effectively.
5. It was observed that the thermal conduction capability significantly improves in polyester resin with rise in BN content in the matrix. With the rise in the BN content, the effective thermal conductivity (k_{eff}) enhances by 18.15% and an enhancement as high as 0.449 W/mK is seen in case of polyester filled with 35 wt% of BN (100 μm). For hybrid composites, the increment in k_{eff} is maximum for 20 wt% of BN for polyester-BN/PH composites.
6. As BN content in the composite increases, the value of k_{eff} also keeps on increasing. It was seen that the results obtained from the empirical correlations are in good agreement with the measured values up to a filler concentration of about 20 wt% after which the BN particles start showing a percolation behavior at this weight fraction (20%) which when results in a sudden jump in the thermal conductivity. This point is called the percolation threshold. The percolation threshold is the critical concentration at which BN particles start

contacting with each other forming conductive networks and hence the actual size of the agglomerates become larger. At this point the thermal conductivity enhancement becomes very high due to which rapid conduction occurs.

Recommendations for Potential Applications

1. The light weight and improved thermally conductive polyester-BN composites can be used for various applications such as printed circuit board substrates.

2. With improved thermal conductivity, improved glass transition temperature, low thermal expansion coefficient and modified properties of mechanical and thermal, the hybrid composites with inorganic and organic reinforcements with polyester as a matrix by using optimal ratios of Boron nitride as well as peanut husk powder, the composites will be applicable for use in laminates of heat engine casings, for heat sink applications, substrates in electronics etc. where the rate of heat transfer should be rapid and maximum.

Acknowledgements This research was supported by Veer Surendra Sai University of Technology, Burla for which we show our earnest gratitude and are very much thankful to our colleagues who gave a helping hand that greatly assisted the research work. We would also like to show our gratitude to the faculties of National Institute of Technology, Rourkela for sharing their sharing their pearls of wisdom with us during the course of this research. Lastly, we would also like to thank our team of research scholars whose research work encouraged us to find the knowledge gap and carry out work in this area.

Competing Interests Author has declared that no competing interests exist.

Disclaimer: This present book chapter is an extended form of previously published work in ICMPC-2018, materialtoday Proceedings, Volume5, Issue9, Part3, Pages19747–19, 753, 2018, https://doi.org/10.1016/j.matpr.2018.06.337and ICMMSE-2019,materialstoday Proceedings, Volume 19, Part 2, 2019, Pages 289–295, https://doi.org/10.1016/j.matpr.2019.06.762.

References

1. Zhu BL, Ma J, Wu J, Yung KC, Xie CS (2010) Study on the properties of the epoxy-matrix composites filled with thermally conductive AlN and BN ceramic particles. J Appl Polym Sci 118(5):2754–2764
2. Yung KC, Wang J, Yue TM (2008) Thermal management for boron nitride filled metal core printed circuit board. J Compos Mater 42(24):173
3. Yu H, Li L, Kido T, Xi G, Xu G, Guo F (2012) Thermal and insulating properties of epoxy/aluminum nitride composites used for thermal interface material. J Appl Polym Sci 124(1):669–677
4. Bae J, Kim W, Cho W, Lee SH (2000) The properties of AlN-filled epoxy molding compounds by the effects of filler size distribution. J Mater Sci 35(23):5907–5913
5. Zhou W, Qi S, An Q, Zhao H, Liu N (2007) Thermal conductivity of boron nitride reinforced polyethylene composites. Mater Res Bull 42(10):1863–1873
6. Wattanakul K, Satasit S (2012) The versatile method to control the orientation of BN particles in thermoset matrix. J Chem Chem Eng 6(9):769
7. Hill RF, Supancic PH (2002) Thermal conductivity of platelet-filled polymer composites. J Am Ceram Soc 85(4):851–857
8. Mishra D, Satapathy A (2012) A Simulation Study on thermal conductivity of glass bead embedded polymeric system. International Journal of Plastics Technology 16(1):24–38
9. Mishra D, Satapathy A (2012) Processing and thermal conductivity characterization of solid micro-spheres filled polymer composites. Adv Mater Res 445:526–529
10. Mishra D, Satapathy A (2013) Development of theoretical model for effective thermal conductivity of glass microsphere filled polymer composites. Plast Polym Technol 2(2):39–47

Dr. Anish Khan is an Associate Professor, at the Centre of Excellence for Advanced Materials Research (CEAMR), Faculty of Science, King Abdulaziz University, Jeddah, Saudi Arabia. He received his Ph.D. from Aligarh Muslim University, India, in 2010. He has research experience working in the field of synthetic polymers and organic–inorganic electrically conducting nano-composites. He completed a Postdoctoral in Electroanalytical Chemistry from the School of Chemical Sciences, Universiti Sains Malaysia (USM) in 2010–2011. He has research and teaching experience and published over 300 research papers in referred international journals. He attended more than 20 international conferences/workshops and has published 60 books, 6 in progress, and 107 book chapters. He has completed around 31 research projects. He served as an Editor and board member of the Journal of Natural Fibers (JNF) and a Member of the American Nano Society. Published two US patents. His fields of specialization are polymer nano-composite/cation-exchanger/chemical sensor/micro biosensor/nanotechnology, application of nano-materials in electroanalytical chemistry, material chemistry, ion-exchange chromatography, and electro-analytical chemistry, dealing with the synthesis, characterization (using different analytical techniques) and derivatization of inorganic ion-exchanger by the incorporation of electrically conducting polymers, preparation, and characterization of hybrid nano-composite materials and their applications, polymeric inorganic cation-exchange materials, electrically conducting polymeric, materials, composite material use as sensors, Green chemistry by remediation of pollution, heavy metal ion selective membrane electrode, and biosensor on the neurotransmitter.

Luminescence of Boron Nitrides

Young-Kuk Kim and Jae-Yong Jung

Abstract Recently, a considerable interest has been focused on 2-dimensional hexagonal boron nitrides due to their prominent optical, thermal, chemical and electronic properties. Especially, h-BN consists of 2-dimensional hexagonal net-like layers of boron and nitrogen atoms and such atomic layers were stacked along normal direction. In contrast to isostructural graphene, h-BN is regarded as a wide band gap material with indirect band gap suitable for host of various impurities with light emitting performance. In particular, it was reported that incorporating foreign atom including rare earth elements can provide intense luminescence on h-BN. Unique properties of h-BN based nanophosphors can provide support on various optical application of the same materials. Here, we reviewed various reports on synthesis and luminescence of doped h-BN. In addition, potential application of the highly luminescent h-BN was also summarized.

Keywords Hexagonal boron nitride · Synthesis of nanophosphors · Luminescence · Rare earth doping

1 Introduction

Boron nitride is a stable refractory material consisting of light elements of boron and nitrogen. Various polymorphs of boron nitride such as cubic, hexagonal, rhombohedral structures are found in nature and high similarity with isoelectronic carbon-based compounds. Among these, hexagonal boron nitride (h-BN) are well-known and have various industrial application due to their thermal and chemical stability, unique mechanical, electronic and optical properties. h-BN consists of 2-dimensional hexagonal net-like layers of boron and nitrogen atoms. Those layers are stacked along

Y.-K. Kim (✉)
Korea Institute of Materials Science (KIMS), #797 Changwondaero, Changwon, Kyungnam, Republic of Korea
e-mail: voice21@kims.re.kr

J.-Y. Jung
Pukyung National University, #45 Yong-Soro, Busan, Republic of Korea

M. Jawaid and A. Khan (eds.), *Multifunctional Boron-Nitride Composites*,
Composites Science and Technology, https://doi.org/10.1007/978-981-99-2866-8_7

direction normal to layer. So, h-BN is isostructural to graphene which is a primary crystal lattice of 2-dimensional materials. In contrast to highly conductive graphene, h-BN is regarded as a wide band gap materials with indirect band gap. Although the typical band gap h-BN is around 5 eV, various different reports on the band gap energy of h-BN are found in the literature covering from 3.6 to 7.1 eV depending on state of materials and characterization tools [1, 2]. As reported by many research groups, h-BN can be exfoliated by various chemical or mechanical processes like graphene and atomistically thin layer can be separated. Completely isolated single atomic layers of h-BN are theoretically expected to transition to be direct band gap at near 6 eV [3]. Owing to their excellent electronic and structural nature, h-BN finds various optical application including photonic devices, phosphors, etc. Here, the nature of luminescence from h-BN and their various application will be reviewed based on previous literatures.

2 Synthesis of Boron Nitrides

Synthesis of boron nitride (BN) has been many studied fairly extensively, broad resemblance of BN carbon in its structural forms, combined with certain physical characteristics, makes BN a material of considerable theoretical interest and of growing importance in nuclear, ceramic, electrical, and lubrication technologies [4–6]. In the synthesis of such as BN, there is a method of synthesizing hexagonal boron nitride (h-BN) by direct nitridation reaction of boron according to the following reaction [7, 8].

$$2B + N_2 = 2BN \quad (1)$$

However, this method needs pure boron requires, heterogeneous reaction, and high temperature during the nitridation [9–11].

Recently, methods such as solid-state reaction [12], hydrothermal [13], precursor sintering [14], and chemical vapor deposition [15] have been proposed to lower the reaction temperature and to synthesize BN in a relatively simple method. In addition to these nitridation methods, a boron compound containing oxygen was used as a source of boron atoms and nitriding of boron precursors with a amine-based compound such as ammonia or ammonium compound. On the other hand, borates such as borax ($Na_2B_4O_7$) and boric acid (H_3BO_3) are heated with a nitrogen-containing amine-based compound such as melamine ($C_3H_6N_6$), ammonia (NH_3), and urea ($(NH_2)2CO$) to form a covalent B-N bond. These reactions are usually performed at temperatures 800–1200 °C for pyrolysis and reaction of precursor compounds [16]. The borax-urea mixture heated in an ammonia gas stream to obtain BN. This is a traditionally utilized synthesis method for preparation of h-BN powders, which has cheap cost, low investment, and simple process as synthesis reaction [17]:

$$Na_2B_4O_7 + 2(NH_2)_2CO \rightarrow 4BN + Na_2O + 4H_2O + 2CO_2 \quad (2)$$

$$Na_2B_4O_7 + 2NH_4Cl + 2NH_3 \rightarrow 4BN + 2NaCl + 7H_2O \quad (3)$$

As inferred from the reaction formulae including presence of various by-product, the prepared boron nitride has poor in purity and in uniformity of particle size. Then, the BN produced with this reaction is suitable for industrial production [18]. Among the suggested methods, h-BN was synthesized from H_3BO_3 and $C_3H_6N_6$ were dissolved in of distilled (D.I) water. After evaporated water at 110 °C, melamine borate adduct (M2BA; $C_3N_6{\cdot}2H_3BO_3$) through intermolecular hydrogen bonding was formed and then dried in an oven 80 °C for 24 h. The dried M2BA was pyrolyzed at various temperature for 1 h in a horizontal tubular furnace under nitrogen atmosphere [19]. The pyrolysis characteristics of M2BA checked with thermogravimetric analysis and differential thermal analysis (TG–DTA) showed an abrupt weight loss at 600–800 °C, which can be attributed to formation of BN. Additory weight loss occurring at around 1500 °C possibly caused by nitrogen desorption from the made BN. The M2BA was pyrolyzed at 900 °C although a small amount of B_2O_3 was monitored with x-ray diffraction (XRD) analysis even after washing with D.I water, most of identified XRD peaks were consistent to those of h-BN. After pyrolysis, the main peak (002) phase of h-BN was observed. However, the full width at half maximum (FWHM) of the peak (002) phase for the pyrolyzed samples was much larger than that of the crystalline h-BN. Thus, the formed phase has the same structure as h-BN, but it was identified as a turbostratic boron nitride (t-BN) rotating with hexagonal layers randomly stacked along the c-axis direction [20, 21]. As the pyrolysis temperature increased, FWHM decreased, but unknown peaks were observed at 1300 °C due to unreacted produced. In addition, boric acid and melamine were prepared in a stoichiometric ratio of 2:1 to synthesize the h-BN nanoparticles by two methods. First (wet method), the compound obtained by putting in D.I water (100 °C) and stirring for 48 h is subjected to filter treatment and calcination at 500 °C for 3 h to form a precursor. Second (dry method), the raw ematerials were mixed in a stoichiometric ratio using an agate mortar and pestle. The obtained mixture was heated at 200 °C for 1 h and 300 °C for 2 h. The alumina crucible containing the precursors prepared by two different methods was placed in an alumina tube furnace and exposed to under nitrogen atmosphere for 90 min in 2.4 kW microwave irradiation [22].

The prepared precursors were in the form of melamine diborate ($C_3H_6N_6{\cdot}2H_3BO_3$) and were changed to metaboric acid (HBO_2, Eq. (4)) and boron oxide (Eqs. (5) and (6)) as shown in the reaction formula below during the calcination process [22]. The changed samples were exposed to microwaves in a nitrogen atmosphere to synthesize crystalline h-BN and the (002) phase, which is the main peaks of h-BN could be showed.

$$C_3H_6N_6 + 2H_3BO_3 \underset{\rightarrow}{\Delta} C_3H_6 \cdot 2H_3BO_3 \quad (4)$$

$$H_3BO_2 \underset{\rightarrow}{\Delta} H_2O + HBO_2 \quad (5)$$

$$2HBO_2 \underset{\rightarrow}{\Delta} H_2O + B_2O_3 \quad (6)$$

The synthesized BN nanoparticles had a size of about 20–30 nm and had a roundly plate shape. There is also a case in which a method capable of synthesizing boron hexagonal crystalline in a short time by exposure to microwaves irradiation using a precursor prepared with boric acid and melamine.

Furthermore, Synthesis of h-BN via solid-state reaction method from boric acid, urea, and lithium slat (Li_2O, LiCl, and $LiNO_3$) are blended in a mortar to form a mixture, and then preheated at 200 °C to form a precursor. The precursor is heat-treated in an ammonia gas atmosphere at 1100 °C to synthesize h-BN. It was shown that lithium ions have excellent crystallinity even at a temperature relatively lower than the conventional synthesis temperature during the synthesis of h-BN and can be used as a catalyst for particle growth [23].

In addition, studies on the synthesis of h-BN by hydrothermal method have been reported [24]. The hydrothermal route is popularly used for synthesis of metal oxides o elemental metal nanoparticles, and research on preparation of superfine nitride powder is still in its infancy. Variety of merits including easy control of synthesis condition, synthesis of nanosized particles with high uniformity and sphericity were already identified [25]. Water or organic solvents can be applied as reaction media to achieve high temperature and high pressure environment in an authoclave. Substance with poor solubility can be dissolved in such high temperature and high pressure environment to form new crystals [26]. H_3BO_3 and the aliphatic amine such as melamine, urea, etc. were dissolved in D.I water. The dissolved mixture of those precursors was transferred to a teflon-lined reaction vessel for high pressure and heated in the convection oven at 150 °C. Here, the reaction condition was maintained for 4 h. Fourier transform infrared spectroscopy (FT-IR) was carried out to obtain information on the chemical composition of the BN nanoparticles (8 s). The obtained absorption spectrum of synthesized BNNPs denote a broad absorption peak assigned to the O–H stretching and N–H bending modes. Then, functional groups with –OH units were identified on the surface of BNNPs through chemical bonds with Boron (B) sites. Rather weaker absorption bands including 1638, 1431 and 1369 cm^{-1} were also detected and attributed to B-N stretching modes. The average size of the synthesized nanoparticles was characterized with a transmission electron microscopy (TEM). Nanosized particles were identified from TEM characterization and the average particle size was estimated to be 11.38 nm [27].

Besides these methods a case of synthesizing h-BN using this chemical vapor deposition (CVD) has also been reported. The h-BN was prepared by the CVD based on a hot-wall reactor and highly reactive precursor vapor-like molecules containing B and N are introduced into the reaction chamber with assistance of a carrier gas, and chemical reactions occur between the raw materials at high temperatures in the reactor. Among the variety of gaseous boron precursor, boron-containing compounds such as BF_3, BCl_3, BBr_3, B_2H_6 and $B(OCH_3)_3$ were popularly utilized. Here, ammonia (NH_3) or nitrogen (N_2) was used as nitrogen-containing molecules. The h-BN powders with high purity and sphericity were prepared by the CVD method, and precise control of various processing factors is required during the synthesis [27]. The h-BN in the form of a monolayer or nanostructure using this CVD process. Studies have also been reported in which these nanostructures are stacked in several layers and presented as h-BN two-dimensional materials applicable to semiconductors and flexible nanoelectronics [28].

Recently, in addition to the synthesis of h-BN powder and thin film growth, research on the synthesis of hexagonal boron nitride nanotubes (BNNT) has been actively conducted using ball milling, CVD, thermal plasma jet, arc discharge, and laser ablation methods [29–33]. Control of the conversion rate from boron and nitrogen sources into BN radicals is essential to efficient synthesis of high quality BNNT. In addition, their synthesis requires appropriate strategies including precursors, processing conditions of the equipment to promote the growth of nanostructures. Nanostructured BNNT powders shows variety of important materials properties such as high mechanical strength, high thermal conductivity, electrically resistance, and piezoelectricity originates from their unique crystal structure and compositions.

3 Luminescence of Boron Nitrides

h-BN is isostructural to graphene and has been comprising one of the most important family of low dimensional materials. Since h-BN can be crystallized in diatomic layers combined with weak Van der Waals interaction in out-of-plane direction and with strong covalent bonding in in-plane direction, h-BN has very high anisotropic nature of structural, chemical and electronic properties. One example of the anisotropy is that h-BN can easily be separated to platelets or extremely thin nanosheets like other 2D materials. In contrast, however, h-BN exhibits a quite different electronic behavior compared with graphene which has delocalized π-electrons and shows a semi-metallic nature without band gap. The in-plane B-N bonding is polar covalent since the difference in electronegativity of boron and nitrogen exceeds 0.4 eV. h-BN is classified as an insulating material with wide band gap, since electrons are localized near nitrogen atoms. Although bulk crystals of h-BN has an indirect band gap, they shows intense deep ultraviolet emission owing to strong exciton-phonon interaction [34]. In addition, thickness-dependent nonlinearities were also found in the h-BN. The significant anisotropy in crystal structure of h-BN provides anisotropy in refractive index. Then, h-BN shows a high birefringence implying direction-dependent light propagation speed. The permittivity which is a square root of refractive index in the optical frequency range is different for different direction both by its magnitude and by its signs. Then, h-BN is known as a natural hyperbolic material showing one or two negative real part of dielectric permittivity. The natural hyperbolic properties of h-BN provides opportunity of broad range of applications in mid infrared frequency including hyperlensing, surface enhanced IR absorption (SEIRA), etc.

Defects in h-BN can also contribute special photonic devices, although defects reduced carrier and phonon lifetimes which is harmful to conventional optical devices. Defects in h-BN sometimes contributes to realize single photonic emitters (SPE) [35]. SPE emits only one photon per excitation cycle and defects in solid materials can be treated as artificial atoms which is confined in host materials with a large band gap encompassing ground and excited states within band gaps without thermal recombinations. SPE can be excited with optical or electrical stimulation. SPE shows polarized emission with highly narrow linewidth (~a few nm) enabling high SNR (signal-to-noise ratio). SPE can be realized for h-BN by various crystalline forms such as single crystals, commercial powders, nanotubes, nanoonions, exfoliated flakes and epitaxial thin films, although no specific principles to control emission wavelength and emitting sites. Defects in h-BN were sometimes introduced by irradiation with oxygen atoms [36]. Detailed study on the nature of defects showed that association of nitrogen vacancy and carbon impurity (V_NC_B) and charged boron vacancy (V_B^-) can be attributed to main source of optical emission. SPE by h-BN can be excited by laser of which energy is lower than band gap of h-BN. So, green and blue light emitting laser are popularly accommodated to excite SPE. As an alternative to such lower-than-band-gap excitation, anti-Stokes or up-conversion excitation by optical pumping the phonon side bands [37].

Defects in h-BN can also be introduced with foreign atoms for luminescence. One of the most important dopant for h-BN is carbon [38]. Carbon doping can modify electrical properties of h-BN and insulating h-BN nanosheets can be transformed to conductors. In addition, the magnetic moments can be provided to non-magnetic h-BN by forming triangular carbon cluster inside h-BN controlled with electrical charging. Carbon doping also modifies optical transparency of h-BN on whole visible range of optical spectrum. Carbon doping with CO utilizes advantages of physoption or chemisorption at defect sites in h-BN and removal of residual oxygen atom in h-BN by forming CO_2. Sometimes, carbon can be incorporated to single crystals during their growth with graphite furnace. Here, the carbon doping can change from colorless or transparent yellow color indicating completion of carbon doping. Carbon doping can introduce shallow and deep levels, which can be manifested by variation in photoluminescence spectra (PL). The zerophonon line (ZPL) was identified by low temperature micro PL spectra of C-doped h-BN excited by 2.41 eV laser [39]. From the variety of optical spectra from micro-PL, the light emission from C-doped h-BN was attributed their origin to collections of carbon-related defects which form well-defined midgap levels enabling the radiative center-to-center recombination.

Oxygen-doping can also modify the electronic structure and optical properties of h-BN [40]. Theoretical simulation of BN nanotubes or nanosheets indicates clear narrowing of band gaps by substitutional doping of oxygen for nitrogen in the lattice of h-BN. Besides, the modulation of optical, electronic and magnetic properties was also expected. From experiment of treating surface of h-BN with oxygen plasma, the optically measured band gap of BN was reduced to be 4.31 eV. Also, absorption at (near) UV regions was also found with oxygen doped BN. Thus, oxygen doping is an efficient tool to modify electronic properties of h-BN. The photoluminescence in visible spectrum is hardly observed with pure h-BN and nitrogen vacancy or oxygen impurities can generate intense photoluminescence at visible ranges due to localized states from defects. The large amount of oxygen doping can reduce impurity-to-impurity distance and forming new bands by delocalization of defect states. Interestingly, excitation with blue lights ($\lambda \sim 450$ nm) made anti-Stokes photoluminescence showing PL at higher energy. Here, electrons can interact with phonons to draw to higher energy states for emission with shorter wavelengths. Since such anti-Stokes PL is rare for conventional inorganic materials, further works are required.

It was reported that doping with rare earth elements (REE) can induce intense luminescence from h-BN. The first report on luminescence from REE-doped h-BN dates back to 1960s [40]. Dzhuzeev et al. introduced various REE to BN and observed visible photoluminescence spectra. Recently, Jung et al. reported intense near UV emission from Ce-doped BN [19]. The Ce doping can enhance the crystallinity of h-BN crystals and efficient stacking of BN monolayers shows reduced interplanar spacing even though Ce with larger ionic radius was incorporated to the lattice. The PL spectra from Ce-doped BN showed a broad spectral feature covering wavelength range from 350 to 550 nm, although most intense spectrum is in ~400 nm. The intensity of PL spectra was saturated with doping level of 0.5 mol% showing mirror image with excitation spectra. The mirror symmetry in excitation spectra and photoluminescence spectra is typically found for defect centers with weak Jahn–Teller interaction. The near UV luminescence from Ce-doped BN was ascribed to 4f–5d transition of Ce^{3+} ions.

In addition, co-doping of Ce^{3+} and Tb^{3+} ions to h-BN shows a dramatic change in PL spectra [41]. Concomitant doping of Ce^{3+} and Tb^{3+} ions showed a quite different PL spectra from that of Ce^{3+} doped h-BN. As the doping concentration of Tb^{3+} increased, the broad PL spectrum at ~400 nm which is a characteristic spectrum from Ce^{3+} doped h-BN was minimized. In contrast, the intensity of sharp and discrete PL spectrum from Tb^{3+} ions are enhanced. Those relative variation of PL intensity by changing relative doping level of two REE can be ascribed to energy transfer between REEs. The transfer efficiency (η_{ET}) approached to unity after saturated doping of both REE ions (Fig. 1).

4 Application of Luminescent Boron Nitrides and Their Composites

As mentioned in the previous section, the h-BN with various morphology has been intensively studied due to its thermal conductivity, high melting point, extreme mechanical hardness, high electrical resistance, and large bandgap energy both theoretically and experimentally [42]. It shows a wide bandgap can be used essentially as a direct bandgap UV laser material. Since the first observation of strong far-ultraviolet exciton emission, in the past ten years, the unique optical and fluorescent properties of h-BN deserve special attention, which makes this material a candidate material for

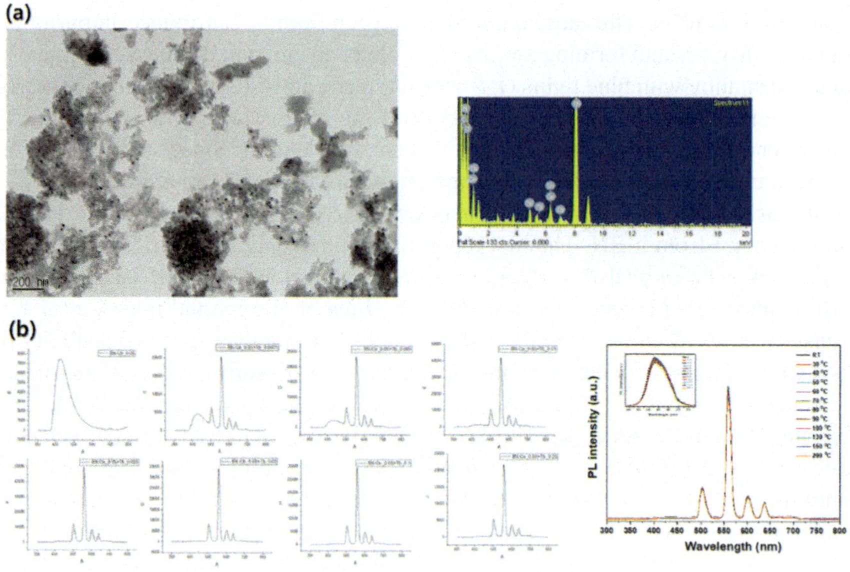

Fig. 1 Nanoparticles of Ce,Tb-codoped h-BN. **a** Microscopic image and compositional spectrum; **b** Variation of luminescence spectra with respect to Tb addition and Temperature

new light-emitting devices and is expected to become the future materials for photovoltaic and optoelectronic applications [43, 44]. Therefore, this section introduces various application of hexagonal boron nitride.

Oxide-based phosphors such as yttrium aluminum garnet (YAG), which are one of the conventionally well-known luminescent materials through rare earth doping by using an oxide material having an excellent crystal structure as a host [45]. Oxide with excellent crystal structure absorbs external energy such as light and heat, transfers energy to the doped rare earth and emits light [36]. In the same manner as the structure of such an oxide-based luminescent material, a study has been reported that suggests that it can be used as a luminescent material by doping with crystalline h-BN as a host and doping with rare earth ions. The preparing a mixture of boric acid and melamine, a small amount of cerium ions is added to form a precursor. When the prepared precursor is subjected to heat treatment in a nitrogen atmosphere of 1200 °C, crystalline hexagonal boron nitride is formed. At this time, a blue-emitting nitride-based phosphor is synthesized by the added cerium ions (Ce^{3+}) [46]. When preparing a precursor by the same experimental method, the phosphor synthesized by heat treatment of a precursor made by adding cerium and terbium (Tb^{3+}) ions co-doping showed emission peaks at 500, 560, 600, and 635 nm. In addition, the energy transfer phenomenon from Ce^{3+} to Tb^{3+} ions showed strong green luminescence intensity (Fig. 1). An opaque colloidal solution containing the Ce^{3+} and Tb^{3+} co-doped BN phosphor was prepared by dispersion in aqueous solution of polyvinylpyrrolidone (PVP). The prepared colloidal solution was deposited on glass substrates by spin coating and transparent thin film of Ce^{3+} and Tb^{3+} co-doped

BN thin film was prepared. Distinction between the bare glass and phosphor-coated glass was very hard by naked eyes under daylight due to its optical transparency. However, intense green colors were shown from deposited thin film under UV light. The same colloidal solution were painted by brush on the surface of a Korean bank note for application as anticounterfeiting. In addition, the phosphors were deposited on the metallic substrate with glass frits for robust attachment. The label painted with BN nanophosphors showed intense green light even after sintering in air at 600 °C for 1 h. In addition, the same glow property was identified even after further heat treatment at higher temperature. Here, essential properties of anti-counterfeiting inks having the concealment in normal condition and possible identification with naked eyes were achieved by the transparency in visible light and clear luminescence under UV irradiation of BN-based nanophosphors. In addition, durability in harsh environment also endorsed potential application of this material for identification of high quality mechanical component working in variety of extreme conditions. In summary, the Ce^{3+} and Tb^{3+} co-doped BN nanoparticles can be regarded as a promising candidate of light emitting label suitable for various anticounterfeiting application [41].

In addition, it can be synthesized with a white light-emitting hexagonal boron nitride phosphor doped with cerium and dysprosium ions. The synthesized white light-emitting phosphor was prepared in the form of a colloidal solution made by mixing with PVP in the same manner as the previously reported phosphor doped with cerium and terbium ions and coated on a glass substrate. It was suggested that change in emission spectra of the coated thin film can be identified only in deep UV light, so that it can be used as a UV detection function. Other phosphors and glass frits (composition: $BaO-ZnO-B_2O_3-SiO_2$) was mixed to make a phosphor in glass (PIG) disk. The produced PIG showed white light emission when applied to a UV-LED chip, and it was shown that phosphors that can be applied to various fields can be synthesized by adding rare earths using boron nitride as a host [47].

Except for the well-known method of synthesizing a luminescent material containing rare earths, another method of luminous hexagonal boron nitride has been reported. The precursor is prepared using boric acid, melamine, and poly (propylene glycol)-poly(ethylene glycol). The prepared precursor is heat-treated under nitrogen atmosphere at 200 °C. The heat-treated precursor was synthesized from porous hexagonal boron nitride and was referred to as activated BN (αBN) [48]. Porous αBN was used as a host and dyes such as rhodamine and fluorescein isothiocyanate (FITC) were added to the empty space, suggesting that it can be used as a light emitting composite material that does not require the addition of rare earth ions. A white light-emitting composite phosphor was presented by adjusting the ratio of the dye to be added, and it was shown that it can be applied as a device by applying it to a blue LED chip.

The BCNO composite is synthesized after heat treatment performed by adding hexamethylenetetramine (carbon source) to the mixture prepared by dissolving boric acid and melamine introduced above in D.I. water. The synthesized BCNO complex is nitrogen vacancy and carbon–oxygen vacancy. It is believed to be emitted by defects. In addition, the luminescence position changes according to the amount of hexamethylenetetramine added, which is reported as a phenomenon caused by carbon related impurity levels. Although there was a change in the luminescence characteristics according to the concentration, a change in the luminescence characteristics according to the synthesis temperature was also observed. Blue shift in the emission spectra was identified with increasing the sintering temperature. Such layer was mainly induced by concentration change and the shift of energy levels of the carbon- and oxygen-related impurities The impurity levels of carbon and oxygen impurities would be formed below the conduction band of −4.1 eV [49]. As the impurity concentration increases, the impurity level will move to the conduction band, causing a red shift in the emission spectrum. As the impurity concentration decreases, the impurity level will move to the valence band, which will cause a blue shift in the emission spectrum. When preparing a precursor with sodium borohydride and urea (CH_4N_2O) for use as a light emitting material using these spatial defects, the powder obtained by ball milling by adding lithium chloride (LiCl) and potassium chloride (KCl) is heat-treated to synthesize a BCNO. Emission spectrum of the same material showed a clear red shift with increasing concentration of the carbon. Such shift in emission spectra can be attributed to incorporation of carbon atoms into the lattice structure and such incorporation of impurities could reduce the band gap or induce defect-related energy levels in the electronic structure of BCNO compounds [50]. The atomic arrangement of BCNO compounds can strongly affect their band gap and this can also make important effect on the optical properties of the materials [51].

5 Summary

Boron nitride is one of the most fascinating materials for optical applications owing to its electronic structures, significant structural anisotropy. So far various experimental works have been endowed to elucidate luminescence mechanism and to develop new application for optical devices including composite materials. In particular, defects in boron nitrides realized a new application, single photonic emitters which emits single photon per excitation cycle. In addition, impurity doping also provide ability of high intensity luminescence as shown from h-BN doped with oxygen and rare earth elements, etc. The luminescence spectra can easily be modified by controlling doped atomic species and amounts of dopants.

In addition, polymeric composites with luminescent h-BN provides an efficient route to transparent fluorescence label suitable for various application including anti-counterfeiting, fingerprint detection. Consolidation h-BN with glass frits results in highly luminescent optical tags with excellent stability against high temperature treatment in oxygen containing environment due to high chemical and thermal stability of boron nitrides.

Along with those rapid developments in optical application of h-BN, fundamental studies on electronic structures are also highly pursued theoretically and experimentally. The results from fundamental studies can contribute to enhance luminescence properties of h-BN and pave an efficient way to high performance optical devices.

Acknowledgements This work was supported by the Nano and Materials Technology Development Project (NRF-2020M3H4A2084780) through National Research Foundation (NRF) of Korea.

References

1. Solozhenko VL, Lazarenko AG, Petitet JP, Kanaev AV (2001) Band gap energy of graphite-ike hexagonal boron nitride. J Phys Chem Solids 62:1331–1334
2. Elias C, Valvin P, Pelini T, Summerfield A, Mellor CJ, Cheng TS, Eaves L, Foxon CT, Beton PhH, Novikov SV, Gil B, Cassabois G (2019) Direct band-gap crossover in epitaxial monolayer boron nitride. Nature Comm 10:2639
3. Caldwell JD, Aharonovich I, Cassabois G, Edgar JH, Gil B, Basov DN (2019) Photonics with hexagonal boron nitride. Nat Mater 4:552–567
4. Mahdizadeh A, Farhadi S, Zabardasti A (2017) Microwave-assisted rapid synthesis of graphene-analogue hexagonal boron nitride (h-BN) nanosheets and their application for the ultrafast and selective adsorption of cationic dyes from aqueous solutions. RSC Adv 7:53984–53995
5. Kobayashi H, Fukuoka A (2017) hexagonal boron nitride for adsorption of saccharides. J Phys Chem C 121:17332–17338
6. Lipp A, Schwetz KA, Hunold K (1989) State of the art hexagonal boron nitride: and applications* fabrication, properties. J Eur Ceram Soc 5:3–9
7. Tay RY (2018) Chemical vapor deposition growth and characterization of two-dimensional hexagonal boron nitride. Springer Singapore, Singapore
8. Yang P, Prater J, Liu W, Glass J, Davis R (2005) The formation of epitaxial hexagonal boron nitride on nickel substrates. J Elec Mater 34:1558–1564
9. Paszkowicz W, Pelka JB, Knapp M, Szyszko T, Podsiadlo S (2002) Lattice parameters and anisotropic thermal expansion of hexagonal boron nitride in the 10–297.5 K temperature range. Appl Phys A 75:431–435
10. Zhou H, Zhu J, Liu Z, Yan Z, Fan X, Lin J, Wang G, Yan Q, Yu T, Ajayan PM, Tour JM (2014) High thermal conductivity of suspended few-layer hexagonal boron nitride sheets. Nano Res 7:1232–1240
11. Chen Y, Chadderton L, Fitz GJ, Williams JS (1999) A solid-state process for formation of boron nitride nanotubes. Appl Phys Lett 74:2960
12. Wood GL, Janik JF, Visi MZ, Schubert DM, Paine RT (2005) New borate precursors for boron nitride powder synthesis. Chem Mater 17:1855–1859
13. McLean B, Eveleens CA, Mitchell I, Webber GB, Page AJ (2017) Catalytic CVD synthesis of boron nitride and carbon nanomaterials—synergies between experiment and theory. Phys Chem Chem Phys 19:26466–26494
14. Hu S, Wang Y, Liu L (2019) Synthesis of h-BN by the combustion of B/KNO_3/octogen mixture under nitrogen atmosphere. J Chem 2019:1–6
15. Gross P, Höppe HA (2019) Unravelling the urea-route to boron nitride: synthesis and characterization of the crucial reaction intermediate ammonium bis(biureto)borate. Chem Mater 31:8052–8061
16. Hao X, Dong S, Fang W, Zhan J, Li L, Xu X, Jiang M (2014) A novel hydrothermal route to synthesize boron nitride nanocrystals. Inorg Chem Commun 7:592–594

17. Alkoy S, Toy C, Gönül T, Tekin A (1997) Crystallization behavior and characterization of turbostratic boron nitride. J Eur Ceram Soc 17:1415–1422
18. Solozhenko VL, Solozhenko EG (2001) Equation of state of turbostratic boron nitride. High Press Res 21:115–120
19. Jung JY, Baek YK, Lee JG, Kim YD, Cho SH, Kim YK (2018) Appl Phys A 124:628
20. Yasnó JP, Kiminami RHGA (2020) Short time reaction synthesis of nano-hexagonal boron nitride. Adv Powder Technol 31:4436–4443
21. Budak E (2018) Low temperature synthesis of hexagonal boron nitride by solid state reaction in the presence of lithium salts. Ceram Int 44:13161–13164
22. Lian G, Zhang X, Zhu L, Tan M, Cui D, Wang Q (2011) Hydrothermal synthesis of cubic boron nitride microcrystals. Res Chem Intermed 37:369
23. Feng S, Li G (2017) Chapter 4—Hydrothermal and solvothermal syntheses. In: Anonymous modern inorganic synthetic chemistry. Elsevier B.V
24. McMillen CD, Giesber HG, Kolis JW (2008) The hydrothermal synthesis, growth, and optical properties of γ-LiBO 2. J Cryst Growth 310:299–305
25. Kainthola A, Bijalwan K, Negi S, Sharma H, Dwivedi C (2020) Hydrothermal synthesis of highly stable boron nitride nanoparticles. Mater Today: Proc 28:138–140
26. Park J, Choi SH, Zhao J, Song S, Yang W, Kim SM, Kim KK, Lee YH (2016) Thickness-controlled multilayer hexagonal boron nitride film prepared by plasma-enhanced chemical vapor deposition. Curr Appl Phys 16:1229–1235
27. Khan MH, Liu HK, Sun X, Yamauchi Y, Bando Y, Golberg D, Huang Z (2017) Few-atomic-layered hexagonal boron nitride: CVD growth, characterization, and applications. Mater Today (Kidlington, Engl) 20:611–628
28. Kim KK, Lee HS, Lee YH (2018) Synthesis of hexagonal boron nitride heterostructures for 2D van der Waals electronics. Chem Soc Rev 47:6342–6369
29. Zhuang C, Xu H, Li L, Liu Y, Ban C, Liu X (2014) Systematic investigation of the ball milling-annealing growth and electrical properties of boron nitride nanotubes. RSC Adv 6:113415–113423
30. Lee CH, Wang J, Kayatsha VK, Huang JY, Yap YK (2019) Effective growth of boron nitride nanotubes by thermal chemical vapor deposition. Nanotechnology 19:455605
31. Kim M, Lee YH, Oh J, Hong S, Min B, Kim T, Choi S (1996, 2020) Synthesis of boron nitride nanotubes using triple DC thermal plasma reactor with hydrogen injection. Chem Eng J (Lausanne, Switzerland) 395:125148
32. Lee CM, Choi SI, Choi SS, Hong SH (2006) Synthesis of boron nitride nanotubes by arc-jet plasma. Curr Appl Phys 6:166–170
33. Gnoffo PA, Fay CC (2013) Laser vaporization and plume chemistry in a boron nitride nanotube production rig. J Thermophys Heat Transfer 27:369–381
34. Zhong B, Huang X, Wen G, Yu H, Zhang X, Zhang T, Bai H (2011) Large-scale fabrication of boron nitride nanotubes via a facile chemical vapor reaction route and their cathodoluminescence properties. Nanoscale Res Lett 6:36
35. Lopes JMJ (2021) Synthesis of hexagonal boron nitride: from bulk crystals to atomically thin films. Prog Cryst Growth Char Mater 67:100522
36. Fisher M, Caridad JM, Sajid A, Ghaderzadeh S, Ghorbani-Asi M, Gammelgaard L, Bøggild P, Thygesen KS, Krasheninnikov AV, Xiao S, Wubs M, Stenger N (2021) Controlled generation of luminescent centers in hexagonal boron nitride by irradiation engineering. Sci Adv 7:eabe7138
37. Wang Q, Zhang Q, Zhao X, Luo X, Wong CPY, Wang J, Wang D, Venkatesan T, Pennycook SJ, Loh KP, Eda G, Wee ATS (2018) Photoluminescence upconversion by defects in hexagonal boron nitride. Nano Lett 18:6898–6905
38. Liu Z, Xue Q, Zhang T, Tao Y, Ling C, Shan M (2013) Carbon doping of hexagonal boron nitride by using CO molecules. J Phys Chem C 117:9332–9339
39. Koperski M, Vaclavkova D, Watanabe K, Taniguchi T, Novoselov KS, Potemski M (2020) PNAS 117:13214–13219
40. Weng Q, Kvashnin DG, Wang X, Cretu O, Yang Y, Zhou M, Zhang C, Tang D-M, Sorokin PB, Bando Y, Golberg D (2017) Tuning of the optical, electronic, and magnetic properties of boron nitride nanosheets with oxygen doping and functionalization. Adv Mater 29:1700695

41. Dzhuzeev U, Ramazanov P (1969) Photoluminescence spectra of boron nitride activated by certain elements. Sov Phys J 12:890–893
42. Jung JY, Song BK, Kim YK (2019) Tunable color emission of transparent boron nitride nanophosphorstowards anti-counterfeiting application. J Alloys Compnds 791:81–86
43. Wu J, Yin L, Zhang L (2013) Tuning the electronic structure, bandgap energy and photoluminescence properties of hexagonal boron nitride nanosheets via a controllable Ce3+ ions doping. RSC Adv 3:7408
44. Shevitski B, Gilbert SM, Chen CT, Kastl C, Barnard ES, Wong E, Ogletree DF, Watanabe K, Taniguchi T, Zettl A, Aloni S (2019) Blue-light-emitting color centers in high-quality hexagonal boron nitride. Phys Rev B 100:155419
45. Xie R, Li YQ, Hirosaki N, Yamamoto H (2011) Nitride phosphors and solid-state lighting. CRC Press, Boca Raton
46. Ronda CR (2008) Emission and excitation mechanisms of phosphors in luminescence: from theory to applications. Wiley-VCH Verlag GmbH & Co. KgaA
47. Jung JY, Shim Y, Son CS, Kim Y, Hwang D (2021) Boron nitride nanoparticle phosphors for use in transparent films for deep-UV detection and white light-emitting diodes. ACS Appl Nano Mater 4:3529–3536
48. Li J, Lin J, Huang Y, Xu X, Liu Z, Xue Y, Ding X, Luo H, Jin P, Zhang J, Zou J, Tang C (2015) Organic fluorescent dyes supported on activated boron nitride: a promising blue light excited phosphors for high-performance white light-emitting diodes. Sci Rep 5:8492
49. Wang W, Ogi T, Kaihatsu Y, Iskandar F, Okuyama K (2011) Novel rare-earth-free tunable-color-emitting BCNO phosphors. J Mater Chem 21:5183
50. Mazzoni MSC, Nunes RW, Azevedo S, Chacham H (2006) Electronic structure and energetics of $B_xC_yN_z$ layered structures. Phys Rev B 73:073108
51. Zhang X, Lu Z, Liu H, Lin J, Xu X, Meng F, Zhao J, Tang C (2015) Blue emitting BCNO phosphors with high quantum yields. J Mater Chem C 3:3311–3317

Boron Nitride Nanocomposites Used as Hydrogen Storage Material

Md. Mohasin, Md. Zafer Alam, Qasim Ullah, and Salman A. Khan

Abstract Hydrogen is considered a best energy source for the next upcoming generation as an energy storage, carrier and vector. It can be precursor and auxiliary material for a number of industrial process. Although hydrogen has the best potential of satisfying the demand for fuel in the future, there are many obstacles limiting its practical application, and storage of hydrogen is very challenging due to it's very high energy density relative to its weight. Boron Nitride nanocomposites has mostly been reported for the storage of molecular hydrogen by chemisorption and physisorption over a wide range of temperatures and pressure. Different BN nanocomposite has been used for the storage of Hydrogen like BN nanotubes, h-BN, p-BN, Open-ended BN and BN-Clusters. By increasing surface area, pressure and decreasing temperature hydrogen adsorption increases by physisorption. By doping with O and C with B and N element of boron nitride hydrogen adsorption takes by chemisorption. The d-block element like Pt, Pd, Ti, Si, Al are doped with Boron nitrite nanocomposite further enhancement of hydrogen adsorption takes place.

Keywords Nanotubes · Nanocomposites · Hexagonal boron nitride · Hydrogen storage

1 Introduction

The research community has been working hard to address the issues related to energy consumption as a result of the requirement to support a rapidly expanding population. Finding new, efficient fuel sources to replace conventional ones like fossil fuels, which are continually being depleted and generate harmful gases including COx, NOx, and SOx. Hydrogen is a clean burning molecule with high calorific value with zero carbon emission, it is as one of the most advantageous renewable and sustainable energy sources for commercial applications due to its qualities like its availability

Md. Mohasin · Md. Z. Alam · Q. Ullah · S. A. Khan (✉)
Physical Sciences Section (Chemistry), School of Sciences, Maulana Azad National Urdu University, Hyderabad 500032, Telangana, India
e-mail: sahmad_phd@yahoo.co.in

M. Jawaid and A. Khan (eds.), *Multifunctional Boron-Nitride Composites*,
Composites Science and Technology, https://doi.org/10.1007/978-981-99-2866-8_8

from ample resources like photo splitting of water. Because of its effectiveness, lightweight and environmental friendliness, hydrogen is a clean alternative energy carrier for both stationary and mobile applications [1–7]. Hydrogen can also be utilized in the pharmaceutical sector for drug discovery, drug development, medical imaging and spectroscopic technique. In the field of metallurgy, the use of hydrogen in oxy-hydrogen flames for cutting and welding metals as well as for reducing metals from solutions [8]. There are numerous initiatives and studies underway to switch electric vehicles to hydrogen fuel-cell vehicles, which emit no carbon dioxide and can be quickly refuelled in a matter of minutes. Although hydrogen holds out the best chance of meeting future fuel demand, there are numerous barriers preventing it for the practical implementation. The main cause is that hydrogen has a very high energy density relative to its weight, [9] which makes the conventional method of converting hydrogen into liquid hydrogen storage very challenging [10–12]. An efficient, secure, and stable storage media is necessary for the usage of hydrogen. Numerous hydrogen storage systems have been described, and they can be broadly divided into two groups: (a) Chemical Storage Systems and (b) Physical Storage Systems. The chemical hydrogen storage systems have strong hydrogen bonds and large storage densities, they cannot regenerate the storage medium, which makes their practical application challenging. Physical hydrogen storage system has week binding energy and reversible in nature. Hydrogen can be stored in small volumes by raising the pressure up to 700–800 bars, or by lowering the temperature i.e. liquid hydrogen at -253 °C [13]. In other ways Physical hydrogen storage can be done by using metal organic frame woks, chemical hydrides, metal hydrides and zeolites [14]. Additionally, high pressure and large expenses are required for storage of Hydrogen in the liquid phase. So, there is still a need for pertinent studies on how to create safe and effective hydrogen storage. Currently, a lot of research is being done on advanced materials including chemical hydrides, absorbents, and adsorbents as well as upgrading current technology [15]. Nanomaterials including those based on boron and carbon have recently drawn a lot of attention for a variety of uses. Because it is anticipated that they would have special qualities like high temperature stability and high wear resistance material for making cutting tools and abrasives. Since they are expected to have special qualities like high temperature stability and wear resistance for making cutting tools and abrasives [16].

Boron Nitride is a chemical compound contain 1:1 ratio of boron(B) and nitrogen(N) atom, their molecular formula is BN. Covalent bonds hold the in-plane atoms together, whereas the out of plane layers are held together by weak Vander Waals force of attraction and producing anisotropic characteristics.BN offers large band gap around 5.5 eV and low density which is around 2.26 g/cm^3. In addition to offering good thermal conductivity, excellent thermal stability, large surface area, strong electrical resistance, light weight, low dielectric constant and loss tangent and non-toxicity, good microwave transparency and simple machinability. Additionally, it is lubricating, non-reactive, and non-abrasive to molten metals [17, 18]. Boron Nitride(BN) nanocomposites are most significant systems that are receiving attention as effective hydrogen storage materials. It is due to its high oxidation state, high mechanical strength and low density [19]. The ionic nature of of the B-N bond in BN is an interesting characteristic since it induces an increased dipole moment and higher hydrogen adsorption [20]. Because hydrogen's induced dipole moment is sensitive to local electric fields. It is a close analogue to carbon nanotubes in many ways. Storage of Hydrogen of on different Boron Nitrate nanocomposites has been discussed below.

2 Boron Nitride Nanotubes

Boron Nitride Nanotubes have important application for the hydrogen storage. It possesses alternating B and N atoms, with chirality geometry. Boron nitride nanotubes, which are correlate to carbon nanotubes [21], it has physical characteristics that make them appropriate for a wide range of applications [22]. With nearly identical band gaps of 5.5 eV, BNNTs have broad band gap semiconductors that are not affected by the tube's width, helicity, or number of tube walls [23]. These Boron Nitride Nano tube are thermally and chemically more stable [24–26]. The practical use of BNNT in hydrogen storage is very high. BNNT can absorb up to 2.6 wt% hydrogen at roughly 10 MPa at ambient temperature, according to experiments [27]. Jhi and Kwon reported that Boron Nitride Nano Tube had a greater Hydrogen adsorption energy than Carbon Nano Tube [20]. According to Mpourmpakis et al. stated that BNNT are preferred than CNT for applications requiring Hydrogen storage [28].

The structure resembles like bamboo nanotube. Later these were named as Boron Nitride whiskers [29]. First theoretical predictions for the Boron Nitride nanocomposite with tubular structure was done in 1994 [23, 30]. Boron Nitride can be synthesized experimentally via arc-discharge method in 1995 [31]. The majority of research in the years that followed concentrated on the structural characterisation of Boron Nitride Nanotubes after their synthesis. Several techniques, including chemical vapour deposition [32], substitution reaction [33], ball milling [34], and laser ablation [35], were developed and used to synthesize BNNTs.

Here we can discuss about two kinds of BN nanotubes samples, these are shown in figure Fig. 1 given bellow. (a) multiwall nanotubes and (b) bamboo-like nanotubes. Scale bar: 100 nm.

These two BN nanotubes were synthesized by using chemical vapour deposition technique, at which B-N-O precursor was pyrolyzed at around 2000 K under N_2/NH_3 atmospheric condition. By changing the oxygen amounts in the B-N-O scaffold, the shapes of the nanotubes being controlled [27]. Based on Transmission Electron Microscopy (TEM) Characterization, multiwall BN nanotubes have lengths of very few micrometres and a narrow diameter between 10 and 30 nm. Each end of the tube is capped. H_2 can be stored with large extend in the interior cavities of BN nanotubes due to its capped like feature. Bundles are often formed by the bonding of a few nanotubes. Bamboo-like tubes typically have lengths greater than 10 nm and occupy a significantly larger diameter range of 10 to 80 nm in the shape of linked nanobell segments [36].

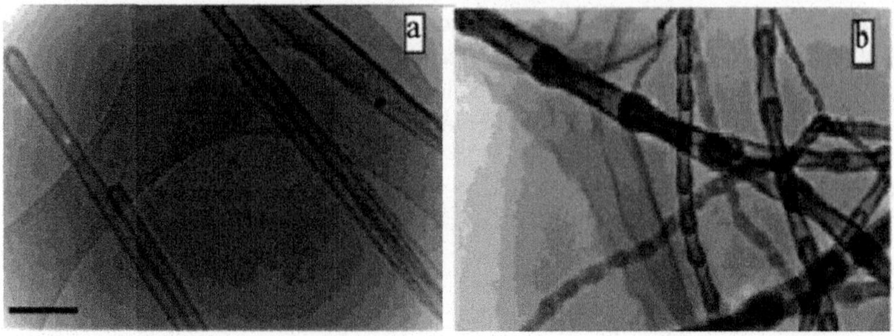

Fig. 1 The morphologies of Boron Nitride nanotubes: (**a**) Multiwall Boron Nitride nanotubes (**b**) bamboo-like Boron nitride nanotubes. Reproduced from [27]. Copyright 2002, JACS, Communication

The amount of Hydrogen adsorption increases uniformly with incrasing pressure, which is measured in Boron Nitride multiwall and bamboo nanotube up to 1.8–2.6 wt% at 10 MPa pressure. This shows that Boron Nitride nanotubes have a significantly improved capacity to adsorb hydrogen than bulk boron nitride powder. Because of the nonporous structure, these boron nitrides have greater adsorption rates can be easily understood by looking at their larger specific surface areas [27].

3 Hexagonal Boron Nitride(h-BN)

Hexagonal Boron nitride(h-BN) is a very promising materials for hydrogen storage and applications due to its distinctive structure, strong chemical and thermal stability, and polar B-N bond. According to theoretical study the additional dipole moment arises due to B-N ionic bond which favour a higher Hydrogen adsorption than its carbon equivalent [20]. Wang et al. first reported that h-BN as a hydrogen absorbent material in 2002,following research based on hydrogen storage material on graphite [37–39]. Hexagonal Boron Nitride based materials are promising for several applications due to their low cost and reversible Hydrogen uptake [40]. After 80 h of ball milling in an Hydrogen atmosphere, nanostructured hexagonal Boron Nitride with lateral sizes of 3 nm absorbed up to 2.6 wt% of hydrogen chemically and physically. The nano material could maintain its flawed nanostructure up to 1173 K, whereas recrystallization of nanostructured graphite upon desorption began at 950 K [41]. Hexagonal Boron Nitride structures with diverse topologies, particularly tubular structures, were used to achieve high hydrogen storage at room temperature. Hexagonal Boron Nitride structures, including hollow spheres with ultra-thin shells [42], collapsed nanotubes [43], nanofibers [44], bamboo-like nanotubes [27] and porous whiskers with micropores or mesopores [45], showed good hydrogen uptake of 2.6–5.6 wt% at room temperature. Porous Hexagonal Boron Nitride with high specific surface area, eleactronic and chemical structure [40, 46–48] have high tendency for hydrogen uptake.

After doing a structural optimization of hexagonal Boron Nitride, it was reported that the optimal bond lengths of B-N is 1.45 A^0 and bond length of N-N/B-B were 2.50 A^0 respectively. These values are consistent with the earlier reported results [49–51].

The various Hydrogen adsorption sites and the optimum arrangement of the BN nanosheet are shown in the figure bellow. In the below given Fig. 2 Hydrogen adsorbed in hexagonal boron nitride system with various Hydrogen molecules at different adsorption sites like. a-(centre), b-(bridge) c-(Boron top), d-(Nitrogen top). (a)0-Hydrogen, (b)1-Hydrogen(c)2-Hydrogen(d)3-Hydrogen(e)4-Hydrogen(f)5-Hydrogen(g)6-Hydrogen(h)7-Hydrogen, (i)8-Hydrogen are filled in the respective hexagonal boron nitride structure.

The hydrogen molecule's bond length is 0.74 A^0 and 3.02 A^0 distances is there between hydrogen molecule and hexagonal Boron Nitride plane from vertically. It can be observed that the centre is the best probable adsorption site out of four available ones in hexagonal boron nitride nanocomposites [52].

When hexagonal boron nitride(h-BN) undergoes functionalization with p-phenylenediamine to form modified h-BN-Ph-NH_2. Then its hydrogen storage capacity increases from 1.26 to 2.71 wt% at 77 K and 80 bar Pressure [53].

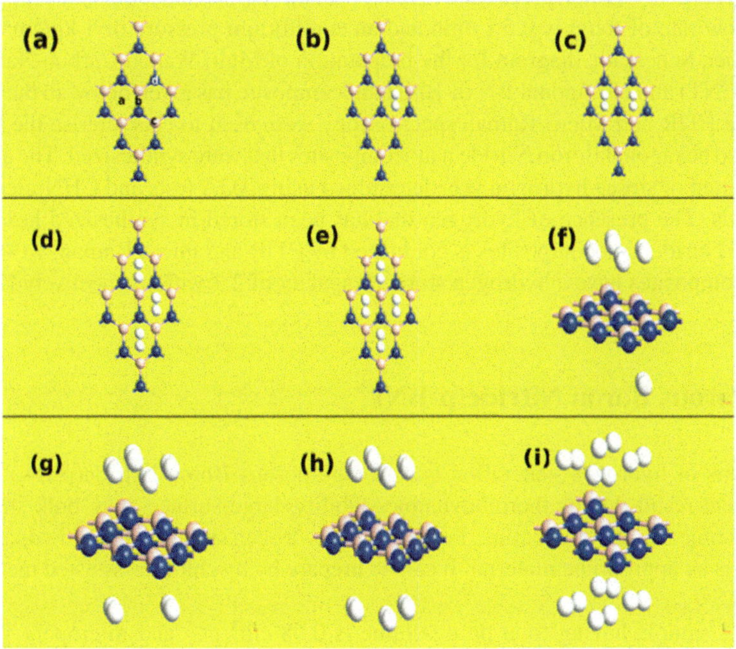

Fig. 2 Different structures of Hydrogen adsorbed h-BN systems with different number of Hydrogen molecules. Reproduced from [52]. Copyright 2021, Surface and Interfaces, Elsevier B.V

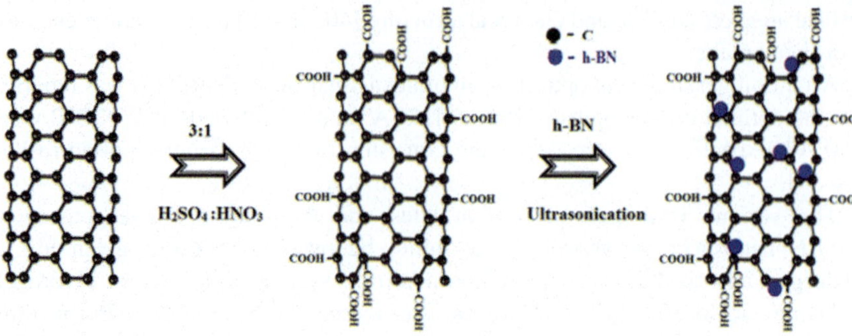

Fig. 3 Schematic diagram for the preparation of Multi Wall Carbon Nano Tube and hexagonal boron nitride nanocomposite. Reproduced from [14]. Copyright 2016, Renewable Energy, Elsevier Ltd.

The hydrogen storage capacity increases at low temperature (cryogenic) and high pressure.

The synthesized of Multi Walled Carbon Nanotube (MWCNT) and hexagonal Boron Nitride (h-BN) nanocomposites were synthesized by the facile ultra-sonication method. For hydrogenation, these nanocomposites are kept in a setup like Sievert's-like hydrogenation. The nanocomposites were kept at 100 °C in vacuum for one hour. The hydrogen was then given time to react with these BN nanocomposites for 15 min at a flow rate of roughly 0.5 L/min and an equilibrium pressure of 1 kg/cm^2 in the chamber. Schematic diagram for the preparation of Multi Walled Carbon Nanotube (MWCNT) and Hexagonal BN (h-BN) nanocomposite has given below in the Fig. 3.

The FTIR and Micro-Raman spectroscopy were used to characterise the hydrogenated hexagonal Boron Nitride nanocomposites that were synthesized. The amount of emitted or stored hydrogen was determined using TGA tests and CHN-elemental analysis. The presence of hydrogen that has been stored in synthesized hexagonal Boron Nitride nanocomposites is confirmed by FTIR and micro Raman. wt% h-BN nanocomposites have a hydrogen storage capacity of 2.3 wt%, respectively [14].

4 Porous Boron Nitride(p-BN)

In terms of hydrogen generation and storage porous Boron Nitride(p-BN) shows excellent result due its thermodynamic stability, large surface area, bulk structure and chemically active in nature. For physical Hydrogen storage and cryo-adsorption, p-BN is an appropriate material. It can be prepare by by chlorine-assisted method at 600 °C.

This sample having total pore volume is 0.68 $cm^3\ g^{-1}$ and micropore volume is 0.38 $cm^3\ g^{-1}$ and specific surface is 960 $m^3\ g^{-1}$, these textural characteristics caused it to reversibly store H_2 is around 1.01 wt% at at -196 °C and 1 bar pressure.

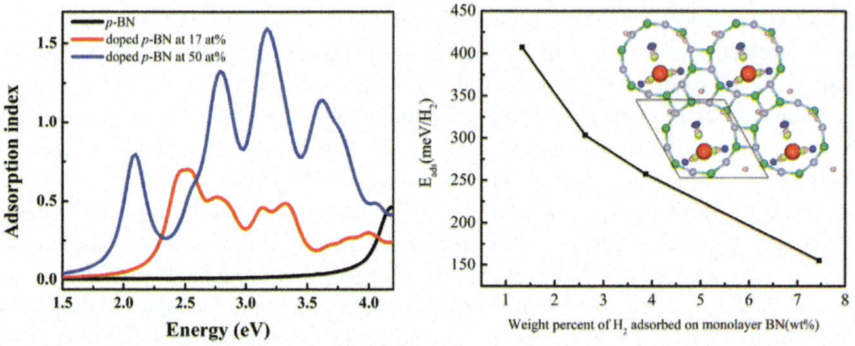

Fig. 4 Calculated optical absorption spectrum of original Fig. 5. The adsorption energy of hydrogen on Li- and C-doped p-BN with different constitute percentages. decorated p-BN and optimized geometry with maximum hydrogen. Reproduced from [54]. Copyright 2015, Journal of Materials Chemistry A, Royal Society of Chemistry

By increasing specific surface area and pressure reversible H_2 uptake increase up to 2.3 wt% at −196 °C and 10 bar pressure. Pore width is another factor that affects the H_2 uptake extent. For example, it was reported that Boron Nitride sponge like having highly porous up to 1.07 $cm^3\ g^{-1}$ to 1900 m^3g^{-1} and having its widths of 1 nm showing reversible storage is 2.57 wt% at −196 °C and 10 bar pressure.

The band gap of the p-BN is 3.98 eV which very high. It can be decrease further by doping with 50% carbon which becomes 1.8 eV band gap. The doped p-BN shows moderate bond gap with strong adsorption of Hydrogen around 5.1 wt% under visible light region [54] (Fig. 4).

5 Hydrogen Storage in Boron Nitride Clusters

Hydrogen Can Be Stored Both in Carbon Nanotube as Well as Boron Nitride (BN) Nanomaterial

LaB_6 and Pd/boron powder were used to synthesise boron nitride (BN) nanomaterials.

LaB_6 and Pd were chosen due to its catalytic properties which enhances its hydrogen storage by chemisorption process. At room temperature, physisorption of hydrogen is challenging due to the hydrogen's weakly bonded state, which prevents large-scale storage. This is true both interstitial position and inside the nanotubes. The hydrogen storage was then examined using differential thermogravimetric analysis.

Boron Nitride (BN) nanotubes and nanocapsules are synthesized by melting Metal and boron particle by arc-melting method. As the liquid solidifies into metal or boride nanoparticles, residual born reacts with nitrogen to form Boron Nitride layers on the surface of Boron Nitride nanoparticles. Due to its electrical insulation, BN fullerene materials are often produced using the arc-discharge process and particular conducting electrodes like HfB_2 and ZrB_2. For the creation of BN nanomaterials, the mixed powder arc melting approach offers two benefits. It is not necessary to use specific electrodes because pressing causes the powder to become conductive. At 600 and 1000 °C, respectively, carbon fullerenes and boron nitride fullerenes succumb to sublimation. In comparison to carbon fullerenes, boron nitride fullerenes store hydrogen gas with less energy and exhibit good thermal stability. Materials made of boron nitride fullerene might be a better choice for H_2 storage.

Tetragonal and hexagonal rings in Born Nitride nanostructures would have energy barriers preventing Hydrogen molecules from passing through them. The energy barrier value in case of C_{60} hexagonal rings is 16 eV at tetragonal rings. In case of $B_{36}N_{36}$ hexagonal rings the energy barrier is 14 eV which smaller value in compare to tetragonal ring which is 27 eV. Hydrogen molecule passes more easily within hexagonal B36N36 rings than C60 hexagonal ring. The comparison shows that, Hydrogen molecule passes through B36N36 hexagonal rings is easier than C_{60} hexagonal ring and $B_{36}N_{36}$ tetragonal rings [55].

Hydrogen molecules were placed into cluster centres, and energies for these molecules were determined in order to examine their stability. According to calculations, the energies of C_{60}, $B_{24}N_{24}$, $B_{36}N_{36}$, and $B_{60}N_{60}$ are respectively 0.58, 20.71, 20.93, and 20.82 eV/mol atom. This study found that the H_2 molecules in $B_{24}N_{24}$, $B_{36}N_{36}$, and $B_{60}N_{60}$ are more stable than the H_2 molecules in C_{60}. In BN cluster $B_{36}N_{36}$ is the most stable cluster. In Fig. 5 Structural model of different molecular cluster is shown below [55] (Fig. 6).

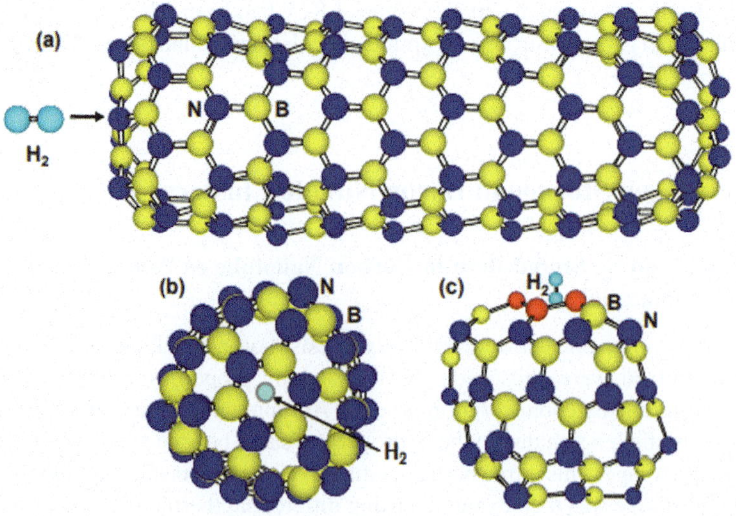

Fig. 5 According to structural model, hydrogen molecules passes through hexagonal Boron Nitride rings of (**a**), (**b**) tip of $B_{99}N_{99}$ nanotube and (**c**) $B_{36}N_{36}$ cluster. Reproduced from [55]. Reproduced from [55]. Copyright 2014, Energies, Enrico Sciubba

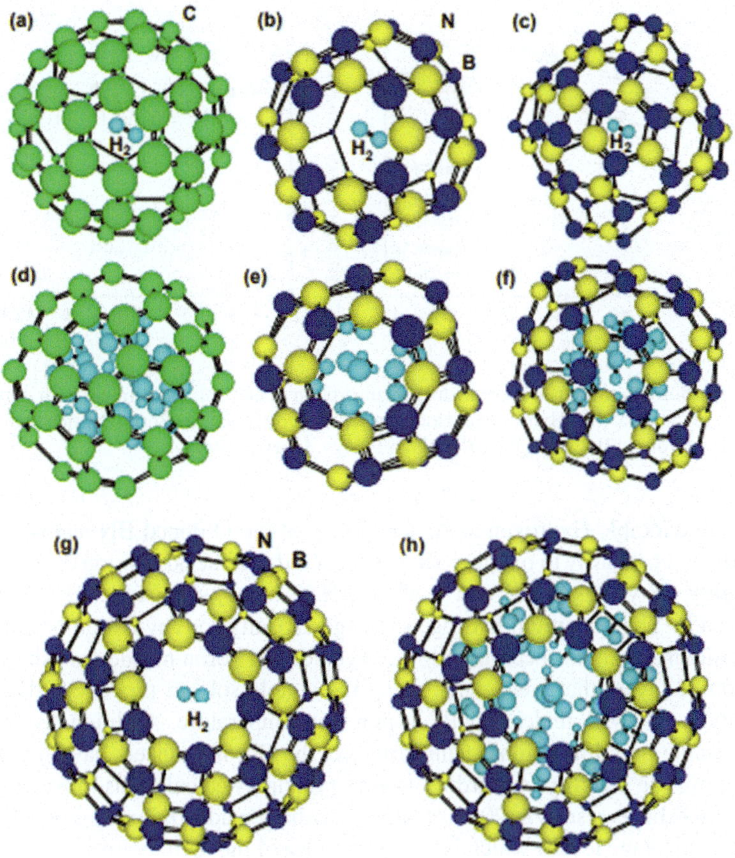

Fig. 6 Model showing the structure of Hydrogen molecule in clusters. The Hydrogen molecule occupied in the centre of (**a**) C_{60} (**b**) $B_{24}N_{24}$ (**c**) $B_{36}N_{36}$ (**d**) 22 Hydrogen molecules occupied inside C_{60} Cluster and 8 atoms are chemisorbed; (**e**) 9 Hydrogen molecules occupied in $B_{24}N_{24}$ cluster and (**f**) 20 molecules occupied in $B_{36}N_{36}$ Cluster (**g**) 1 and (**h**) 38 Hydrogen molecules occupied in $B_{60}N_{6}$. Reproduced from [55]. Copyright 2014, Energies, Enrico Sciubba

6 Doping of Boron Nitride

The purpose of substitutional doping is to replace some of the B or N atoms in BN with a lighter atom. The changing the chemistry and structure of Boron Nitride by doping with different atoms to enhance the hydrogen adsorption on Boron nitride surface. Carbon has simply been offered as a substituent because of its placement in the periodic table opposite B and N. When compared to clean Boron Nitrite Nano Tube, carbon-doped BNNTs have higher hydrogen binding energies. When hydrogen is adsorbed on top of a carbon atom at a boron site, it shows a donor defect level, whereas when it is adsorbed at a nitrogen site, it shows an acceptor defect level [56, 57].

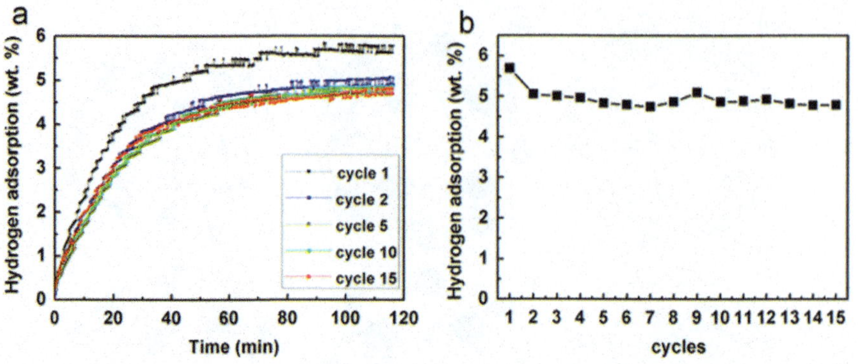

Fig. 7 (**a**) Boron nitride nanosheet hydrogen adsorption at ambient temperature with a hydrogen pressure of 5 MPa. (**b**) Hydrogen adsorption on cyclically stable Boron Nitride nanosheets. Reproduced from [47]. Copyright 2014, Nano Energy, Elsevier Ltd.

The remarkable Hydrogen storage capacity of the O-doped BN nanosheets can be attributed primarily to oxygen. They discovered that the adsorption energy, which is roughly 4.8 versus 8.7 kJ mol^{-1} when BN is doped with oxygen, is about 80% higher [47]. The maximum hydrogen storage capacity for any Boron Nitride nano composite material has been found in oxygen doped boron nitride nanosheets with 2–6 atomic layers that were made using the simple sol–gel process. This storage capacity is 5.7 wt% under 5 MPa at ambient temperature. Additionally, because of the stable two-dimensional nanostructure, the BN nanosheets have remarkable storage cycling stability. According to first-principles calculations, oxygen doping of BN nanosheets results in 20–80% higher Hydrogen adsorption energies on BN at different coverage levels, which is the main cause of the high hydrogen storage [47] (Fig. 7).

Doping through substitution by C or O is effective. Kumar et al. used dispersion-corrected DFT to confirm this. Additionally, they took into account co-doping by Carbon and Oxygen and discovered that the modified Boron Nitride would be significantly more effective at holding Hydrogen than its O-, C-, and 2C-doped equivalents Which is shown below in the Fig. 8 [58].

White spheres stand in for the hydrogen molecule.

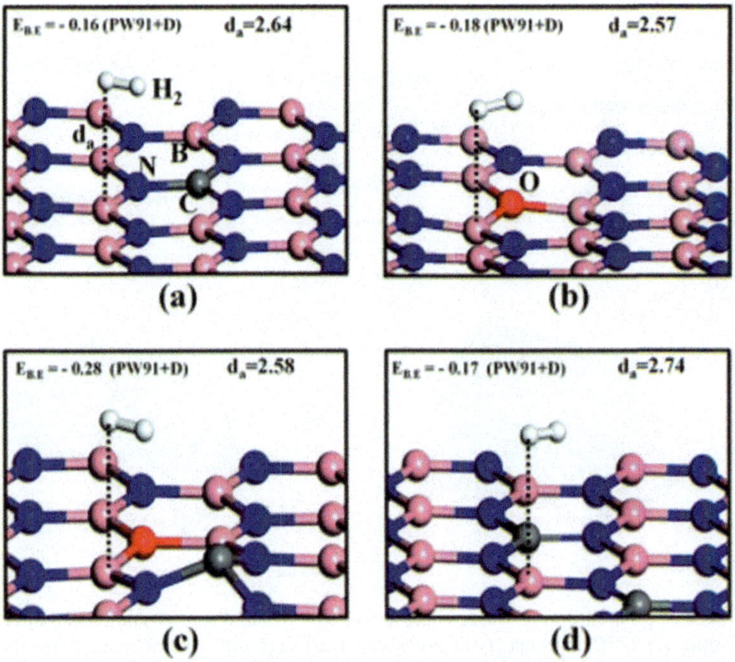

Fig. 8 Single hydrogen adsorption geometry. (**a**) C-doped Boron Nitride, (**b**) O-doped Boron Nitride, (**c**) C-and O-doped Boron Nitride (**d**) 2C-doped Boron Nitride Sheets. The grey, red, blue and pink spheres refer to C,O,N and B atoms. White spheres stand in for the hydrogen molecule. The Pure hexagonal Boron Nitride binding energy is −0.15 eV. Reproduced from [58]. Copyright 2014, Journal of Materials Chemistry A, Royal Society of Chemistry

7 Functionalization of Boron Nitride

Functionalization is an additional method for enhancing the Hydrogen storage capacity of Boron nitride, which is called as decoration of boron nitride nanocomposite surface with heteroelement. Lithium is a best element for the functionalization of boron nitride. By using ab initio first-principles calculations, Wang et al. estimated that each Li atom adorning a chain of atoms in boron nitride would hold up to six molecules of hydrogen [59].

Zhang et al. made a similar statement, stating that Li-decorated porous Boron Nitride can only theoretically adsorb up to 7.5 weight percent of Hydrogen and that the average adsorption energy is around −15.4 kJ mol^{-} [54]. The authors noted that a synergetic effect for Hydrogen physisorption would result from combining high porosity and functionalization. Shahsavari and Zhao's reported that computational studies show that doping pillared Boron Nitride with Li+ increases Hydrogen uptake by 16% at 27 °C and 100 pressures, the adsorption energy of Li+ is predicted to be −166.2 kJ mol^{-1}, resulting in Hydrogen uptake of approximately 10.3 wt% compared to 8.7 wt% for dopant-free Boron Nitride [60] (Fig. 9).

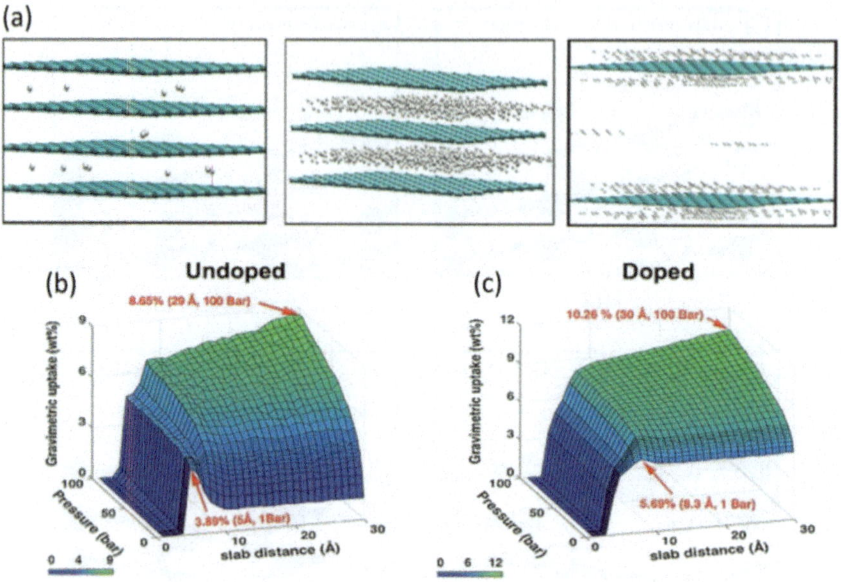

Fig. 9 (**a**) The pillared Boron Nitride sheet model with 3 slab distances and related to Hydrogen concentrations at 27 °C and 1 bar Hydrogen is shown above that the central image shows the greater hydrogen concentration for the ideal slab distance. (**b**) Hydrogen uptake of Undoped Boron Nitride at at 27 °C is shown in a 3D graphic as a function of different slab distances and Hydrogen pressures. (**c**) Li+-doped pillared BN is shown in a 3D plot of the Hydrogen uptake at 27 °C as a function of different Hydrogen slab distances and pressures. Reproduced from [59]. Copyright 2018, Small, Communication

As a potential sorption site for Boron Nitride, sodium and other alkali derivatives were recommended as an alternative method. This approach would be to use such alkali elements such as Li_2F, ONa and OLi to modify a Boron Nitride structure at the molecular basis [59, 61]. This would result in additional heterosites and structural defects, which would increase Hydrogen uptake. The utilisation of $LiBH_4$ as a chemical Hydrogen storage substance that can withstand heating up to 300 °C. It was found that adding 10 wt% of $LiBH_4$ to a sample of BN increased its specific surface area (from 70 to 122) and enhanced Hydrogen absorption with 2.3 wt% at 100 °C and 1 bar [62].

Modification of a Boron Nitride structure using Li_2F, ONa, and OLi at the molecular level.

It is well known that transition metals have a specific affinity for hydrogen. Therefore, they could be used as decorative component on Boron Nitride surface. Computational techniques were used to study some of them like Pt, Pd, Rh, Ni, Co, Fe, Mn, Cr, V, Ti and Sc metals.

If Boron nitride nanotube are doped with Pt metal, Tang et al. reported that hydrogen storage might even increase to 4.2 wt% [43]. The Pt-doped BNNT has a lower energy gap, which suggests that the doping improves the conductivity of the material. The Pt atom in these materials can be seen sticking out from the side-wall. To enable the adsorption of gas, the d orbitals of platinum are exposed to the outside [63]. Defects cause the hydrogen molecule's chemical dissociation to become exothermic and allow the energy barrier of dissociation to be noticeably lowered, which could partially explain the high level of hydrogen chemisorption on BN nanotubes. As seen in Fig. 3a, initially physically adsorption of hydrogen takes place onto the Pt atom with a negligibly low adsorption energy of 0.023 eV. The H–H bond length is approximately 0.77 A^0, while the H-Pt distance is approximately 2.24 A^0. The second hydrogen molecule adsorbed chemically to the Pt-doped Boron Nitride nanotube after overcoming an energy barrier which is shown in Fig. 10. Pink, Blue and White spheres represent the Boron, Nitrogen and Hydrogen atoms [64].

When Pt and Pd atoms are doped with Boron Nitride nano sheet and their hydrogen adsorption properties measured by first-principle approach. According to the result, Pt and Pd atoms showing binding energies around 5.028 and 4.113 eV, prefer to be located on top of the N atom. The nitrogen and transition metal atoms come together to form a strong chemical bond [65].

The Ni and Rh doped with Boron nitride. It was shown that Ni atoms favour top sites over nitrogen, while Rh atoms chose hollow sites over the hexagonal Boron Nitride sheet. In a Ni-doped BN sheet, Nitrogen and Ni metal form a strong chemical bond. It was discovered that the bonds between the boron and nitrogen atoms and Rh atoms mix in BN sheets that have been doped with Rh. Both Ni and Rh atoms have the chemical capacity to bind up to three hydrogen molecules, and as the number of hydrogen molecules increases, so does the distance between the metal atom and the Boron Nitride sheet increases [66].

Titanium is of major interest, primarily because of its low price with effective result for the Hydrogen adsorption. In their study of Ti-doped Boron Nitride nanotubes

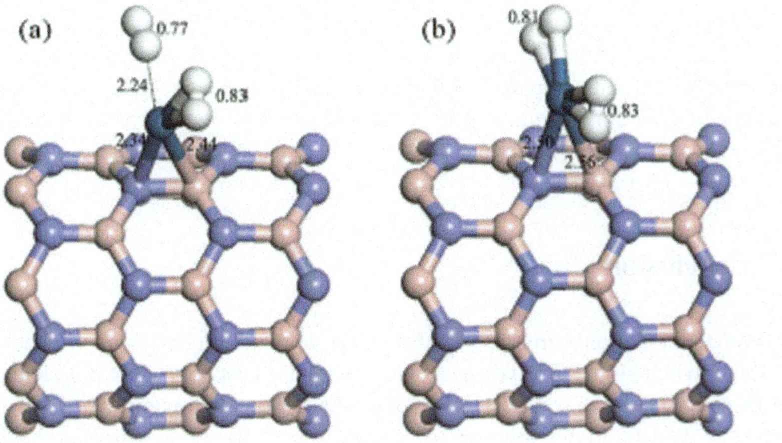

Fig. 10 Adsorption of hydrogen on Pt-doped Boron Nitride nanotubes. Reproduced from [64] Copyright 2006, The Journal of chemical physics, AIP

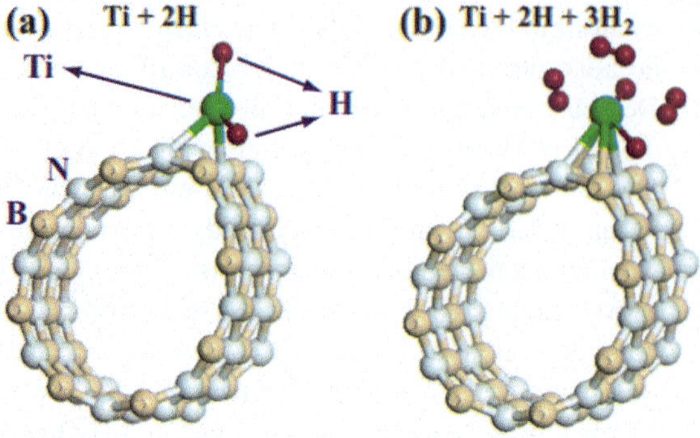

Fig. 11 Hydrogen adsorption on Ti-doped Single walled Boron Nitride nanotube. Reproduced from [68]. Copyright 2006, Physical Review, The American Physical Society

with B-N defects, Mananghaya et al. reported that each Ti atom could hold seven molecules of H_2, indicating a reversible Hydrogen uptake of 7.17 wt% at ambient temperature [67]. The B-N defects would improve Ti's ability to be chemically functionalized, improving its storage capacity of H_2.

Ti atoms can functionalize single-walled boron nitrite nanotubes and each Ti atom is capable of binding up to four hydrogen molecules. One hydrogen molecule is dissociated and the remaining three are molecule are absorbed on the Ti doped boron nitride nanotubes. As a result, the system's hydrogen storage capacity ranges between 3.9 and 5.7 wt%.Hydrogen adsorption takes place in Ti-doped Single walled boron nitride nanotubes which is shown given below Fig. 4a. The initial Hydrogen molecule splits into two H atom. (b) Then three more Hydrogen molecules are absorbed near Ti metal [68] (Fig. 11).

Highly aluminium-doped zigzag and armchair Boron Nitride Nano Tubes have the capacity to adsorb up to 9.4 and 8.6 weight percent of hydrogen, respectively, in an ideal situation [69]. A small increase in the H_2 adsorption at ambient temperature was observed when compared to uncoated nanofibers around 1.5 versus 2 wt% H_2. Similar to titanium, cerium would be capable of holding 7 Hydrogen molecules, giving it a theoretical Hydrogen uptake of 5.68 wt% (Fig. 12) [70].

8 Conclusion

As global population increases rapidly, energy demands increases simultaneously. In order to switch biofuel to Hydrogen energy which is best option due it is clean, green and renewable energy sources. It is challenging to store hydrogen energy due its have very low density, to get solution for this Boron Nitride nanocomposite have wide application to store the hydrogen on its surface by physisorption and chemisorption process. By increasing pressure and decreasing the temperature physisorption takes place, its also depends upon the nature and shape of Boron nitride nanocomposite.

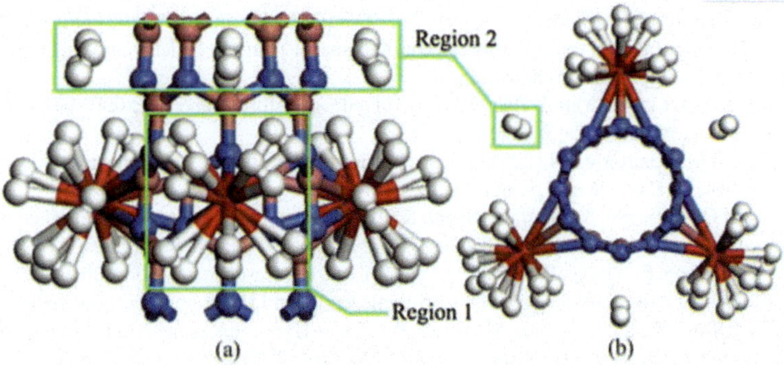

Fig. 12 Boron Nitride nanotube functionalized by cerium metal. The red, grey, pink and blue spheres represent the Ce, C, B and N atoms. **a** Top View and **b** A side view of the atomic structure with the maximum Hydrogen molecules adsorbed for the Ce atoms according to the optimum design. The fine images of regions 1 and 2, which represent the regions near and far from the Ce atoms, are shown in two insets. Hydrogen molecules are oriented similarly in region 1. Hydrogen bond axis is directed toward the N atom in region 2. Reproduced from [70]. Copyright 2006, International Journal of Hydrogen Energy, Elsevier Ltd.

Weather it have more porous or more specific surface area. Chemisorption takes place on the surface of boron nitride by doping with oxygen or carbon in the place of N and B which having nearly equal in size. By doing modifying boron nitride with different d-block metal like Pt, Pd, Li, Ti and Cs hydrogen adsorption further increases.

References

1. Joseph J, Sivasankarapillai VS, Nikazar S, Shanawaz MS, Rahdar A, Lin H et al (2020) 'Borophene and boron fullerene materials in hydrogen storage: opportunities and challenges'. Chemsuschem 13:3754–3765
2. Muthu RN, Rajashabala S, Kannan R (2016) 'Hexagonal boron nitride (h-BN) nanoparticles decorated multi-walled carbon nanotubes (MWCNT) for hydrogen storage'. Renew Energy 85:387–394
3. Wang X, Maeda K, Thomas A, Takanabe K, Xin G, Carlsson JM, Antonietti M (2009) A metal-free polymeric photocatalyst for hydrogen production from water under visible light. Nat Mater 8:76–80
4. Ni M, Leung DY, Leung MK, Sumathy KJFPT (2006) An overview of hydrogen production from biomass. Fuel Process Technol 87:461–472
5. Zhang H, Lin G, Chen J (2010) Evaluation and calculation on the efficiency of a water electrolysis system for hydrogen production. Int J Hydrog Energy 35:10851–10858
6. Holladay JD, Hu J, King DL, Wang Y (2009) An overview of hydrogen production technologies. Catal today 139:244–260
7. Sakintuna B, Lamari-Darkrim F, Hirscher M (2007) Metal hydride materials for solid hydrogen storage: a review. International journal of hydrogen energy 32:1121–1140

8. Okolie JA, Patra BR, Mukherjee A, Nanda S, Dalai AK, Kozinski JA (2021) Futuristic applications of hydrogen in energy, biorefining, aerospace, pharmaceuticals and metallurgy. Int J Hydrogen Energy 46:8885–8905
9. Liu Z, Xue Q, Ling C, Yan Z, Zheng J (2013) Hydrogen storage and release by bending carbon nanotubes. Comput Mater Sci 68:121–126
10. Momirlan M, Veziroglu TN (2005) The properties of hydrogen as fuel tomorrow in sustainable energy system for a cleaner planet. Int J Hydrogen Energy 30:795–802
11. Nicoletti G, Arcuri N, Nicoletti G, Bruno R (2015) A technical and environmental comparison between hydrogen and some fossil fuels. Energy Convers Manage 89:205–213
12. Chen X, Wang L, Zhang W, Zhang J, Yuan Y (2017) Ca-decorated borophene as potential candidates for hydrogen storage: a first-principle study. Int J Hydrogen Energy 42:20036–20045
13. Klebanoff L, Keller J, Fronk M, Scott P (2016) Hydrogen conversion technologies and automotive applications. Hydrogen storage technology. CRC Press, pp 64–95
14. Muthu RN, Rajashabala S, Kannan R (2016) Hexagonal boron nitride (h-BN) nanoparticles decorated multi-walled carbon nanotubes (MWCNT) for hydrogen storage. Renew Energy 85:387–394
15. Doğan EE (2020) Hydrogen production and its storage from solar energy. Adv Mater Sci 20:14–25
16. Masjedi-Arani M, Salavati-Niasari M (2017) Novel synthesis of Zn_2GeO_4/graphene nanocomposite for enhanced electrochemical hydrogen storage performance. Int J Hydrogen Energy 42:17184–17191
17. Wu J, Han WQ, Walukiewicz W, Ager JW, Shan W, Haller EE, Zettl A (2004) Raman spectroscopy and time-resolved photoluminescence of BN and B x C y N z nanotubes. Nano Lett 4:647–650
18. Zhong W, Wang S, Li J, Bechelany MC, Ghisleni R, Rossignol F, Balan C, Chartier T, Bernard S, Miele P (2013) Design of carbon fiber reinforced boron nitride matrix composites by vacuum-assisted polyborazylene transfer molding and pyrolysis. J Eur Ceram Soc 33:2979–2992
19. Weng Q, Wang X, Zhi C (2013) Boron nitride porous microbelts for hydrogen storage. ACS Nano 7:1558–1565
20. Jhi SH, Kwon YK (2004) Hydrogen adsorption on boron nitride nanotubes: a path to room-temperature hydrogen storage. Phys Rev B 69:245407
21. Iijima S (1991) Helical microtubules of graphitic carbon. Nature 354, no. 6348: 56–58.
22. Golberg D, Bando Y, Tang CC, Zhi CY (2007) Boron nitride nanotubes. Adv Mater 19:2413–2432
23. Blase X, Rubio A, Louie SG, Cohen ML (1994) Stability and band gap constancy of boron nitride nanotubes. EPL (Europhys Lett) 28:335
24. Loiseau A, Willaime F, Demoncy N, Hug G, Pascard H (1996) Boron nitride nanotubes with reduced numbers of layers synthesized by arc discharge. Phys Rev Lett 76:4737
25. Bengu E, Marks LD (2001) Single-walled BN nanostructures. Phys Rev Lett 86:2385
26. Nirmala V, Kolandaivel P (2007) Structure and electronic properties of armchair boron nitride nanotubes. J Mol Struct (Thoechem) 817:137–145
27. Ma R, Bando Y, Zhu H, Sato T, Xu C, Wu D (2002) Hydrogen uptake in boron nitride nanotubes at room temperature. J Am Chem Soc 124:7672–7673
28. Mpourmpakis G, Froudakis GE (2007) Why boron nitride nanotubes are preferable to carbon nanotubes for hydrogen storage?: An ab initio theoretical study. Catal Today 120:341–345
29. Ishii T, Sato T, Sekikawa Y, Iwata M (1981) Growth of whiskers of hexagonal boron nitride. J Cryst Growth 52:285–289
30.] Rubio A, Corkill JL, Cohen ML (1994) Theory of graphitic boron nitride nanotubes. Phys Rev 49:5081
31. Chopra NG, Luyken RJ, Cherrey K, Crespi VH, Cohen ML, Louie SG, Zettl A (1995) Boron nitride nanotubes. Science 269:966–967
32. Lourie OR, Jones CR, Bartlett BM, Gibbons PC, Ruoff RS, Buhro WE (2000) CVD growth of boron nitride nanotubes. Chem Mater 12:1808–1810

33. Han W, Bando Y, Kurashima K, Sato T (1998) Synthesis of boron nitride nanotubes from carbon nanotubes by a substitution reaction. Appl Phys Lett 73:3085–3087
34. Chen Y, Gerald JF, Williams JS, Willis P (1999) Mechanochemical synthesis of boron nitride nanotubes. Mater Sci Forum 312:173–178
35. Golberg D, Bando Y, Eremets M, Takemura K, Kurashima K, Yusa H (1996) Nanotubes in boron nitride laser heated at high pressure. Appl Phys Lett 69:2045–2047
36. Ma X, Wang EG, Tilley RD, Jefferson DA, Zhou W (2000) Size-controlled short nanobells: growth and formation mechanism. Appl Phys Lett 77(25):4136–4138
37. Orimo S, Majer G, Fukunaga T, Züttel A, Schlapbach L, Fujii H (1999) Hydrogen in the mechanically prepared nanostructured graphite. Appl Phys Lett 75:3093–3095
38. Fukunaga T, Itoh K, Orimo S, Aoki M, Fujii H (2001) Location of deuterium atoms absorbed in nanocrystalline graphite prepared by mechanical alloying. J Alloy Compd 327(1–2):224–229
39. Orimo S, Matsushima T, Fujii H, Fukunaga T, Majer G (2001) Hydrogen desorption property of mechanically prepared nanostructured graphite. J Appl Phys 90(3):1545–1549
40. Weng Q, Wang X, Zhi C, Bando Y, Golberg D (2013) Boron nitride porous microbelts for hydrogen storage. ACS Nano 7:1558–1565
41. Wang P, Orimo S, Matsushima T, Fujii H, Majer G (2002) Hydrogen in mechanically prepared nanostructured h-BN: a critical comparison with that in nanostructured graphite. Appl Phys Lett 80(2):318–320
42. Lian G, Zhang X, Zhang S, Liu D, Cui D, Wang Q (2012) Controlled fabrication of ultrathin-shell BN hollow spheres with excellent performance in hydrogen storage and wastewater treatment. Energy Environ Sci 5(5):7072–7080
43. Tang C, Bando Y, Ding X, Qi S, Golberg D (2002) Catalyzed collapse and enhanced hydrogen storage of BN nanotubes. J Am Chem Soc 124:14550–14551
44. Ma R, Bando Y, Sato T, Golberg D, Zhu H, Xu C, Wu D (2002) Synthesis of boron nitride nanofibers and measurement of their hydrogen uptake capacity. Appl Phys Lett 81:5225–5227
45. Li J, Lin J, Xu X, Zhang X, Xue Y, Mi J, Mo Z, Fan Y, Hu L, Yang X, Zhang J (2013) Porous boron nitride with a high surface area: hydrogen storage and water treatment. Nanotechnology 24:155603
46. Weng Q, Wang X, Wang X, Liu D, Jiang X, Zhi C, Bando Y, Golberg D (2013) Preparation and hydrogen sorption performances of BCNO porous microbelts with ultra-narrow and tunable pore widths. Chemistry–An Asian J 8:2936–2939
47. Lei W, Zhang H, Wu Y, Zhang B, Liu D, Qin S, Liu Z, Liu L, Ma Y, Chen Y (2014) Oxygen-doped boron nitride nanosheets with excellent performance in hydrogen storage. Nano Energy 6:219–224
48. Portehault D, Giordano C, Gervais C, Senkovska I, Kaskel S, Sanchez C, Antonietti M (2010) High-surface-area nanoporous boron carbon nitrides for hydrogen storage. Adv Func Mater 20:1827–1833
49. Chettri B, Patra PK, Vu TV, Nguyen CQ, Yaya A, Obodo KO, Tran NT, Laref A, Rai DP (2021) Induced ferromagnetism in bilayer hexagonal Boron Nitride (h-BN) on vacancy defects at B and N sites. Physica E 126:114436
50. Topsakal M, Aktürk E, Ciraci SJ (2009) First-principles study of two-and one-dimensional honeycomb structures of boron nitride. Phys Rev B 79:115442
51. Ryou J, Park J, Hong S (2017) Investigations of vacancy structures related to their growth in h-BN sheet. Nanoscale Res Lett 12:1–6
52. Chettri B, Patra PK, Hieu NN, Rai DP (2021) Hexagonal boron nitride (h-BN) nanosheet as a potential hydrogen adsorption material: a density functional theory (DFT) study. Surf Interfaces 24:101043
53. Doğan M, Selek A, Turhan O, Kızılduman BK, Bicil Z (2021) Different functional groups functionalized hexagonal boron nitride (h-BN) nanoparticles and multi-walled carbon nanotubes (MWCNT) for hydrogen storage. Fuel 303:121335
54. Zhang H, Tong CJ, Zhang Y, Zhang YN, Liu LM (2015) Porous BN for hydrogen generation and storage. J Mater Chem A 3(18):9632–9637
55. Oku T (2014) Hydrogen storage in boron nitride and carbon nanomaterials. Energies 8:319–337

56. Baierle RJ, Piquini P, Schmidt TM, Fazzio A (2006) Hydrogen adsorption on carbon-doped boron nitride nanotube. J Phys Chem B 110:21184–21188
57. Li F, Xia Y, Zhao M, Liu X, Huang B, Ji Y, Song C (2006) Theoretical study of hydrogen atom adsorbed on carbon-doped BN nanotubes. Phys Lett A 357:369–373
58. Kumar EM, Sinthika S, Thapa R (2015) First principles guide to tune h-BN nanostructures as superior light-element-based hydrogen storage materials: role of the bond exchange spillover mechanism. J Mater Chem A 3(1):304–313
59. Liu Y, Liu W, Wang R, Hao L, Jiao W (2014) Hydrogen storage using Na-decorated graphyne and its boron nitride analog. Int J Hydrogen Energy 39(24):12757–12764
60. Shahsavari R, Zhao S (2018) Merger of energetic affinity and optimal geometry provides new class of boron nitride based sorbents with unprecedented hydrogen storage capacity. Small 14:1702863
61. Naqvi SR, Rao GS, Luo W, Ahuja R (2017) Hussain T "Hexagonal boron nitride (h-BN) sheets decorated with OLi, ONa, and Li2F molecules for enhanced energy storage." ChemPhysChem 18:513–518
62. Muthu RN, Rajashabala S, Kannan R (2017) Hydrogen storage performance of lithium borohydride decorated activated hexagonal boron nitride nanocomposite for fuel cell applications. Int J Hydrog Energy 42:15586–15596
63. Li XM, Tian WQ, Huang XR, Sun CC, Jiang L (2009) Adsorption of hydrogen on novel Pt-doped BN nanotube: a density functional theory study. J Mol Struct (Thoechem) 901:103–109
64. Wu X, Yang JL, Zeng XC (2006) Adsorption of hydrogen molecules on the platinum-doped boron nitride nanotubes. J Chem Phys 125:044704
65. Ren J, Zhang N, Zhang H, Peng X (2015) First-principles study of hydrogen storage on Pt (Pd)-doped boron nitride sheet. Struct Chem 26:731–738
66. Venkataramanan NS, Khazaei M, Sahara R, Mizuseki H, Kawazoe Y (2009) First-principles study of hydrogen storage over Ni and Rh doped BN sheets. Chem Phys 359:173–178
67. Mananghaya M, Yu D, Santos GN (2016) Hydrogen adsorption on boron nitride nanotubes functionalized with transition metals. Int J Hydrog Energy 41:13531–13539
68. Durgun E, Jang YR, Ciraci S (2007) Hydrogen storage capacity of Ti-doped boron-nitride and B/Be-substituted carbon nanotubes. Phys Rev B 76:073413
69. Noura M, Rahdar A, Taimoory SM, Hayward JJ, Sadraei SI, Trant JF (2020) A theoretical first principles computational investigation into the potential of aluminum-doped boron nitride nanotubes for hydrogen storage. Int J Hydrogen Energy 45:11176–11189
70. Zhang ZW, Zheng WT, Jiang Q (2012) Hydrogen adsorption on Ce/BNNT systems: a DFT study. Int J Hydrogen Energy 37:5090–5099

The manufacturer's authorised representative in the EU is Springer Nature Customer Service Centre GmbH, Europaplatz 3, 69115 Heidelberg, Germany. If you have any concerns regarding our products, please contact ProductSafety@springernature.com

Printed and bound by CPI Group (UK) Ltd, Croydon, CR0 4YY
01/10/2026
02225877-0013